Multiresonator-Based Chipless RFID

Stevan Preradovic · Nemai Chandra Karmakar

Multiresonator-Based Chipless RFID

Barcode of the Future

Springer

Stevan Preradovic
Nitero
Endeavour Hills, Victoria 3802
Australia
stevan.preradovic@ieee.org

Nemai Chandra Karmakar
Department of Electrical &
Computer Systems Engineering
Monash University
Melbourne, Victoria 3800
Australia
nemai.karmakar@monash.edu

ISBN 978-1-4614-2094-1 e-ISBN 978-1-4614-2095-8
DOI 10.1007/978-1-4614-2095-8
Springer New York Dordrecht Heidelberg London

Library of Congress Control Number: 2011943083

Printed on acid-free paper

Springer is part of Springer Science+Business Media (www.springer.com)

"This book is dedicated to the memory of Radivoj Preradovic, a loving and supportive father, husband and son."

"To my beloved parents, Haridhan Karmakar and Rajubala Karmakar"

Foreword

Radio frequency identification (RFID) has been experiencing tremendous growth for the last two decades. In recent years, adoption of RFID into the mainstream identification and tagging systems has been accelerated due to the patronisation of the largest retail chains such as Wal-Mart of USA, K-Mart and Myer in Australia and the US Department of Defence. According to IDTechEx, the market volume of RFID exceeded \$5 billion in 2009, and it is estimated that it will surpass \$25 billion in 2018. RFID has numerous applications in tracking items, supply chain management and logistics, and original part manufacturing and pharmaceuticals. RFID has huge potential applications in many areas due to its flexibility, capability to operate non-line of sight and high data carrying capacity. However, RFID could not be exploited in its full potentials in low-cost item tagging. The reasons are: (a) the end users are not adequately learned of the technology. They are too concerned about their return on investment; and (b) the price of an RFID tag is too high compared to that for an optical barcode. The application-specific integrated circuit (ASIC) chip in the conventional tags makes the tag expensive. This bottleneck has hindered deployment of RFIDs in low-cost item tagging. In recent years, a few research groups have been working on chipless RFID tags. The chipless RFID tags do not contain any ASIC, hence can be printed on the items to be tagged. This book presents the chipless RFID tag and the tag reader that were developed at Monash University. The absence of the chip enables the tag to be competitive in price with the optical barcode. Therefore, the chipless RFID has the potential to replace trillions of barcodes printed each year. The only commercially viable chipless RFID tag is the radio frequency surface acoustic wave (SAW) tag. SAW tags need special materials hence are not printable. The book has presented the design and development of a fully printable multi-resonator-based chipless tag and a dedicated RFID reader unit to read the tag. In the introductory part, the book provides the reader with the most recent technical background and technological breakthrough in chipless RFID development. The book clearly shows the need for a fully printable chipless RFID tag and a powerful and efficient RFID reader for reading the tag. The reader needs to be smart and "agile" enough to read and process the data encoded by the chipless tag which does not use conventional modulation techniques (such as amplitude and phase modulation). In the second part of the book, the design details of the basic building blocks of the chipless tag are presented. The building blocks of the chipless tag are a planar disc loaded monopole antenna and an asymmetrical coupled spiral multi-resonator. The integration of the tag is developed in both microstrip lines and co-planar waveguides (CPWs). Tags up to 35 bits were designed and developed. The chipless tag encodes data via the "spectral signature" technique. Three generations of RFID readers were developed. Generation 1 is an amplitude-only detection of the chipless tag at 2.45 GHz ISM band. Generation 2 is an amplitude and phase detection of the tag at the same frequency. Generation 3 is an ultra-wide band (UWB) reader operating from 5 to 10.7 GHz. This reader is capable of reading a higher number of data bits from the chipless tag. System level design and implementation of the reader are presented in the book. These chipless tag and reader developments are unique in the sense that to the best of my knowledge no other research group has reported such a complete development. Therefore, the *Multiresonator-Based Chipless RFID—Barcode of the Future* is a timely publication. The book covers a broad spectrum of topics: the historical perspective and comprehensive review of chipless RFID tags and modern RFID readers and their architectures, development of a chipless RFID tag by integrating planar spiral resonators and UWB monopole antennas in microstrip and CPW technology, development of directional RFID reader antennas, microwave and UWB transceivers for the RFID reader, chipless RFID tag field trials and also the integration of the chipless tag and RFID reader into a fully operational system. This book includes a comprehensive coverage of the chipless RFID tag and reader design. The book will be a fantastic resource for engineers and researchers working in the RFID industry and postgraduate students who doing research in both active and passive RF and microwave design.

Palmerston North, New Zealand Subhas C. Mukhopadhyay

Preface

Radio frequency identification (RFID) is a modern wireless data transmission and reception technique for applications including automatic identification, asset tracking and security surveillance. As barcodes and other means of identification and asset tracking are inadequate for recent demands, RFID technology has attracted interest for applications such as logistics, supply chain management, asset tracking and security access control. With the continuous development of RFID technology, we can see that new applications of RFID have been emerging from all wakes of our day-to-day lives. The vast majority of RFID transponders (or tags) are usually composed of an antenna and integrated circuit (IC). The IC performs essential data processing tasks. For passive RFID tags, ICs are powered by the energy emitted from the interrogation signals of RFID readers. RFID transponders which use on-board power supplies (such as batteries) are called active RFID tags. Passive RFID tags are low in cost with the expense of short reading range (up to 3 m). Active RFID tags need on-board batteries, have higher cost and can be read even over 100 m. The most expensive component of an RFID tag is its application-specific integrated circuit (ASIC). The cost of RFID limits their potential for the replacement of trillions of barcodes each year. According to the market analysis of IDTechEx, chipless RFID will occupy more than 60% of the total RFID market by 2018 (in 2009 total market value of RFID exceeded $5 billion). Therefore, researchers have been developing chipless RFID tags that can be made much cheaper and comparable to the optical barcodes.

Scientific and technological development has a trade-off between the cost and performance. Chipless RFID is going through the various development phases to maximise the trade-off. Chipless tags offer a low-cost tagging solution at the expense of significant technical challenges. The challenges are non-conventional data encoding techniques, reading distances, collision of proximity tags and printing techniques. All these factors may result in a reduced number of bits that can be encoded in a chipless RFID tag when compared to that for a chipped RFID tag. Since chipless RFID tags have no ASIC chip, data encoding and decoding, frequency of operation and reading methods are very different and unique. In this book, we present a comprehensive review of the latest reported chipless RFID circuits presented in peer-reviewed journals and conferences and open resources such as the internet. The chipless RFID tags are classified based on how they encode data. We distinguish between two types of chipless tags: time domain reflectometry-based and spectral signature-based tags. The chipless RFID tags are classified based on how they encode data. The TDR-based tag encodes data in the time domain by creating a train of echo pulses backscattered to the reader when interrogated. The spectral signature-based tag encodes data in the frequency domain by filtering out certain frequencies using resonant structures. As both types of tags have their pros and cons, it is important to decide which tag is suitable for which application. In this book, we focus on the design and optimisation of spectral signature-based chipless RFID.

A novel RFID system composed of a chipless RFID tag and an associated reader is reported in the book. The chipless tag is a fully passive microwave circuit and uses spectral signatures for data encoding. The tag consists of a multi-resonator coupled to transmitting and receiving antennas. To accommodate multiple bits, the tag operates over the ultra wideband (UWB) frequency spectrum. UWB antennas are used to receive the interrogation signal sent from the reader and transmit the signal back to the reader after performing modulation of the frequency spectra with the multi-resonator. Modulation is performed in both amplitude and phase of the spectrum. A chipless tag up to 35 bits, which operate over 3–7 GHz band, has been designed.

After the successful design of the chipless tag, three prototype readers have been developed. The Gen1 reader was designed to validate the chipless RFID concept using 6-bit chipless tag reading based on amplitude-only detection in S-band; the Gen2 reader is an upgraded version of Gen1 with both amplitude and phase detection capability; and the Gen3 reader is a UWB reader capable of reading up to 35-bits in the UWB band. The integrated reader is a complete system with analog RF and digital control sections, RS232 interface and a PC loaded with graphical user interface (GUI) and software protocol. Both the hardware and software design of the RFID reader and field trials of the designed chipless RFID system have been validated in the real-world environment. Investigation into reader antenna systems in order to enhance reading range has also been performed. A read range up to 70 cm was achieved.

The unique features of the developed chipless RFID system are: (a) a low cost, fully printable tag and (b) a dedicated chipless RFID tag reader. The importance of this concept lies in the fact that chipless RFID tags become comparable to barcodes in terms of the substrate material used and the cost of fabrication. The main application of this chipless RFID system is in short-range tagging of extremely low-cost items such as the Australian polymer banknote. The chipless tag can also be used in other applications such as hand tags for appeals, library access cards, etc.

So why would this book be of interest to a reader? Well the vast majority of RFID books describe the conventional types of RFID systems (active, semi-active and passive) and they focus on the design of the rectifier circuit for the tag, antenna design and improving reading range. This book, on the other hand, presents a completely unique research finding on a chipless RFID tag technology—a multi-resonator-based paradigm chipless RFID tag and its dedicated reader. The main hindrance of mass deployment of the RFID technology is the cost of the tag. Various efforts are envisaged to reduce the cost of the conventional tag. They are designed with reduced number of silicon transistors required in the RFID tag chipset and reduced cost of the IC design process. However, these approaches will not bring the RFID technology near to the optical barcodes with regard to cost. Our approach is to get rid of the silicon chip in the tag completely and fully eliminate the above two items (ASIC and relevant processing) that determine the cost and reading range of a RFID system. To the best of the authors' knowledge, no single book has been published on chipless RFID. The book features the detailed design of a chipless RFID system with a comprehensive overview of other chipless RFID systems as well. Therefore, the RFID community will be tremendously benefited knowing the new technology in details.

Since this book presents novel RFID design pathways and encoding techniques (such as chipless RFID) which provide solutions for low cost and robust tagging, we hope that this book will be welcomed by the scientific community and create a significant impact in future RFID development. Since the majority of RFID systems found on the market are HF (13.56 MHz) and UHF (around 900 MHz) RFID systems, it is possible that an illusion could be created among researchers that nothing else besides system level design and performance characterisation of implemented RFID systems is left for engineers/researchers in the RFID field. However, this is far from the truth and this book is intended to present a comprehensive review of chipless RFID technology that is on the market but also prototyped by research institutions around the world using unconventional data encoding techniques, circuits and operating frequencies to tackle some of the issues that current chipped RFID tags have which prevent it from completely replacing the optical barcode: cost below 1 cent, operating in high temperatures, full printability and zero power requirements and power supply.

The findings, design and measurements presented in this book have been the result of a 4-year PhD research project conducted by Dr. Stevan Preradovic under the supervision of Dr. Nemai Karmakar at the Department of Electrical & Computer Systems Engineering at Monash University in Melbourne, Australia. The project was financed by the Australian Research Council Discovery Grant "*Chipless RFID for barcode replacement.*" The outcomes of the PhD research project have been also presented at the ARC Major Grant Expo 2008 held in Canberra as one of the top ten projects predicted to have a significant impact on society. The chipless RFID tag has been patented as an international patent "RF Transponder." The authors would also like to acknowledge the support of Dr. Sushim Roy and Dr. Isaac Balbin to the development of the chipless RFID technology presented to this book.

The book will be of primary interest to RF/microwave engineers specialised in RFID design, transceiver design, antenna design and passive filter design. However, engineers involved in mixed signal design and embedded system design will find this book interesting as well since the book covers topics in passive and active RF/microwave circuit design along with some digital circuit design.

Endeavour Hills, VIC, Australia — Stevan Preradovic
Melbourne, VIC, Australia — Nemai Chandra Karmakar

Contents

Abbreviations

1D	One dimensional
2D	Two dimensional
3D	Three dimensional
A	Ampere
AC	Alternating current
ADC	Analog to digital converter
ADS	Advanced design system
ARC	Australian Research Council
ASIC	Application-specific integrated circuit
ASK	Amplitude shift keying
AUD	Australian dollar
BP	Battery powered
BPF	Band pass filter
BPSK	Binary phase shift keying
BW	Bandwidth
CAD	Computer-aided design
cm	Centimetre
CPU	Central processing unit
CPW	Co-planar waveguide
CS	Chip select
CST	Computer simulation technology
CW	Continuous wave
DAC	Digital to analog converter
dB	Decibel
dBi	Gain expressed in dB with respect to an isotropic radiator
DC	Direct current
DGS	Defected ground structure
DSP	Digital signal processing/processor

EAS	Electronic article surveillance
ECSE	Electrical and computer systems engineering
EIRP	Equivalent isotropic radiated power
EM	Electromagnetic
FCC	Federal communications commission
FPAA	Field programmable analog array
FPGA	Field programmable gate array
FSS	Frequency selective surface
Gen1	Generation 1
Gen2	Generation 2
GHz	Gigahertz
GS	Gigasample
GUI	Graphical user interface
H	Henry
HF	High frequency
IC	Integrated circuit
IDT	Interdigital transducer
IF	Intermediate frequency
IL	Insertion loss
ISM	Industrial scientific and medical
kg	Kilogram
KHz	Kilohertz
LED	Light emitting diode
LNA	Low noise amplifier
LO	Local oscillator
LOS	Line-of-sight
LPDA	Log periodic dipole antenna
mA	Milliampere
MAG	Magnitude
MHz	Megahertz
MIT	Massachusetts Institute of Technology
mm	Millimetre
MoM	Method of moments
mV	Millivolts
NLIS	National Livestock Information System
nH	Nanohenry
PA	Power amplifier
PC	Personal computer
PCB	Printed circuit board
pF	Picofarad

PSK	Phase shift keying
PNA	Performance network analyser
QF	Quality factor
RCS	Radar cross section
RF	Radio frequency
RFID	Radio frequency identification
RISC	Reduced instruction set computer
RL	Return loss
RLC	Resistance–inductor–capacitor
RTB	RF-to-baseband
Rx	Receiver
SAW	Surface acoustic wave
SDMA	Space division multiple access
s	Second
sin	Sine
SMA	Sub-miniature type A
TDMA	Time division multiple access
TDR	Time domain reflectometry
TFTC	Thin film transistor circuits
Tx	Transmitter
UHF	Ultra-high frequency
USA	United States of America
UWB	Ultra-wide band
V	Volt
VCO	Voltage controlled oscillator
VNA	Vector network analyser
W	Watts
WORM	Write once read many
YIG	Yttrium iron garnet

Chapter 1
Introduction

Abstract This chapter focuses on the basic concepts of radio frequency identification (RFID) and RFID systems. The reader is introduced to a top level view of an RFID system comprising the RFID tag, RFID reader, and middleware. The current and emerging applications of RFID are presented. Limitations of barcodes due to non-line-of-sight operation and reading range have opened the door to RFID as an enabling technology for RFID. However, tag prices are not yet comparable with the low cost price of barcodes. Chipless RFID is proposed as the solution for low cost item tagging. A multiresonator-based chipless RFID tag and associated RFID reader block diagrams and concepts are presented.

1.1 Radio Frequency Identification Chapter

Radio frequency identification (RFID) is a wireless data capturing technique, which utilizes radio frequency (RF) waves for automatic identification of objects. RFID relies on RF waves for data transmission between the data-carrying device called the RFID transponder and the interrogator [1, 2].

A typical RFID system is shown in Fig. 1.1. An RFID system consists of three major components: a reader or integrator, which sends the interrogation signals to an RFID transponder that is to be identified; an RFID tag or transponder, which contains the identification code; and a middleware software, which maintains the interface and the software protocol to encode and decode the identification data from the reader into a mainframe or a personal computer (PC). The RFID reader can only read tags within the reader's interrogation zone. The reader is most commonly connected to a host computer, which performs additional signal processing and has a display of the tag's identity [3]. The host computer can also be connected via Internet for global connectivity/networking.

RFID was first proposed by Stockman [4] in his landmark paper "Communication by Means of Reflected Power" in 1948. Stockman advocates that by alternating the load of the tag antenna, it is possible to vary the amount of reflected power

S. Preradovic and N.C. Karmakar, *Multiresonator-Based Chipless RFID: Barcode of the Future*, DOI 10.1007/978-1-4614-2095-8_1,

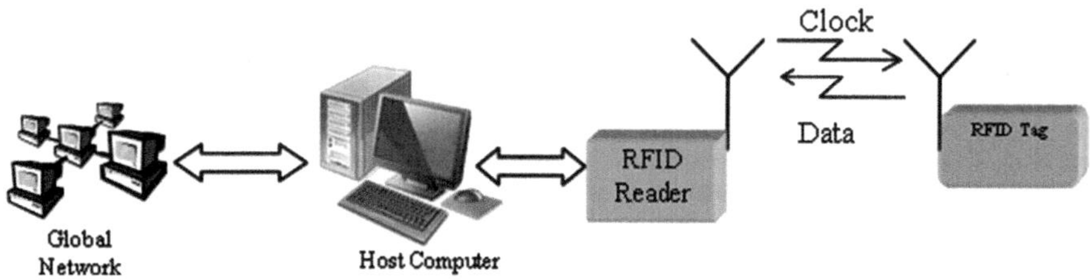

Fig. 1.1 Block diagram of a typical RFID system

(also called "antenna load modulation"), and therefore perform modulation. This new form of wireless technology is known as RFID today. Since then researchers and engineers have been working on developing low cost RFID systems. In the following section, the applications of RFID technology are presented.

1.2 RFID Applications

One of the biggest advantages of RFID technology in comparison to other identification technologies is the option of automated identification and tracking. This is possible due to the non-line-of-sight identification that comes with the use of RF waves. Hence, RFID has found broad areas of applications, which include supply chain management and logistics, electronic article surveillance (EAS), document tracking, luggage handling, vehicle tagging, security and access control, livestock tagging, automobile industry and sports, etc.

With the implementation of RFID into supply chain management came the automation of supply chain management operations and protocols. Thus, reducing the annual profit losses due to human error (scanning of goods manually) and running out of stocks in retail outlets (RFID enables early detection of depleted goods and resources). The first and major retail chain to mandate and successfully implement RFID in their supply chain management was Wal-Mart [5].

Another major application of RFID is EAS, which utilizes ultra low cost 1-bit RFID tags to prevent the theft of goods and articles [6]. Although EAS provides almost no flexibility in terms of tag programmability, their low cost and reasonable detection range (0.5 m) has made them extremely popular in retail chains.

Knowledge management is extremely important to any organization, hence the use of RFID in document tagging has increased with the development of RFID technology. Properly tracking documents using RFID provides cost savings, reduced effort to locate documents, redo or rewrite a document, increases confidentiality, and prevents losing of documents [7].

Another major application of RFID (especially in Australia) is the tagging of livestock. Australia has its comprehensive National Livestock Information System (NLIS) [8]. Under the umbrella of NLIS, every animal and pet in Australia must be e-tagged.

Recent outbreak of mad cow disease and ban of imported beef from some countries, have forced Australia to adopt the NLIS in order to, in case of a disease outbreak, perform quick and effective infected animal quarantines.

Although RFID has found applications in many areas, there is no single RFID system that meets the criteria for all applications. The reason for this is that some features are required in some applications, but not in others. As for example, some applications require short range (up to 1.5 m) low cost tags (luggage tagging), while others require long range (over 20 m) and more robust tags (expensive equipment and vehicle tagging). Hence, we can conclude that the design of an RFID system or choice of an RFID system is determined by the application for which it will be used.

The following section will focus on the limitations of barcodes that have enabled the emergence of RFID as the leading identification technology in the future.

1.3 Limitations of Barcodes and Emergence of RFID as an Enabling Technology

Barcode labels have been used to track items and stocks for sometime after their inception in the early 1970s. Though barcodes are printed in marks and spaces and very cheap to implement, they impose undeniable obstacles in terms of their short range readability and un-automated tracking. These limitations are costing large corporations millions of dollars per annum [9].

The growing tendency today is to replace the barcodes with RFID tags, which have unique ID codes for individual items that can be read at a longer distance. Hence, the obstacles of reading range and automation would be solved using RFID. The only reason why RFID tags have not replaced the barcode is the price of the tag. The cost of an existing RFID tag is still much higher when compared to the price of the barcode. A comprehensive overview of available RFID tags in the market is presented in ref. [10].

The main cost of an RFID tag comes from the chip embedded as the information carrying and processing device in the tag. Huge investments and investigations focus on lowering the price of the RFID chip. As a result, the price of the RFID tag is getting lower and lower every year [11, 12]. However, the price of the RFID tag is still not competitive when compared with the cost of the barcode. The recent development of chipless tags without silicon integrated circuits (ICs) has lowered the cost of the tags comparable to that of the barcode. However, the technology is still at conceptual level.

The tagging of documents and large volumes of paper-/plastic-based items, such as postage stamps, tickets, note bills, and envelopes, is a problem due to the relatively high price of the tag. Chipless RFID tags that can be printed on paper and plastic using conductive ink could prove to be a viable and economical solution. However, the design of a fully functional printable chipless RFID tag has not been reported so far. Hence, a clearly stated gap in RFID technology can be found for tagging the above-mentioned low cost items. A need for fully printable and chipless

RFID technology has been identified as a low cost, efficient, secure, and reliable solution [13]. This project reports the design of fully printable chipless tags based on passive microwave resonators and antennas.

The following section provides a brief outlook of reported chipless RFID systems that can be found in open literature today.

1.4 Chipless RFID Systems

The concept of chipless RFID tags [14] seems to be a promising solution for low cost item tagging. In order to minimize cost, tags are made fully printable and without ICs. Encoding data without an IC is done by two chipless tag encoding schemes: time domain reflectometry (TDR) and spectral signatures.

So far, the only commercially successful chipless RFID system is developed by RFSAW© [15]. The RFSAW tag is based on surface acoustic waves (SAWs) and utilizes TDR encoding scheme. Although SAW tags are fully functional and could well replace the chipped tags, they do not provide a fully printable solution due to their piezoelectric nature, which could not be printed on banknotes, postage stamps, or other paper-/plastic-based items. Hence, the problem of having a robust tagging system for paper/plastic products remains open for research and development.

Printable TDR-based chipless tags have been reported in refs. [16, 17]. These tags encode data using a train of precisely delayed backscattered pulses of the interrogation signal (1 ns pulse) by using multiple capacitive loadings at particular points of the microstrip line. Each data bit needs a delay line, which significantly increases the size of the tag. The drawbacks of this technology are the amount of bits that can be encoded, the size of the tags, and the amount of spectrum used.

Fully printable chipless RFID tags, which use spectral signature encoding, have been reported using space-filling curves [18] and capacitively tuned dipoles [19]. The space-filling curves require considerable layout modifications for data encoding. The capacitively tuned dipoles could have undesired parasitic mutual coupling effects and large layout areas restricting efficient data encoding. This project will present a fully printable chipless RFID system based on multiresonators and cross-polarized ultra wideband (UWB) monopole antennas. The tag's unique ID is encoded as the spectral signatures of the resonators. The main differences between the proposed system and the ones reported in refs. [18, 19] are as follows: we encode data in both amplitude and phase and the operation is not based on radar cross-section (RCS) backscattering. The proposed system works on retransmission of the interrogation signal with the encoded unique spectral ID. The received and transmitted signals are cross-polarized in order to achieve good isolation between the two. Due to the robust design based on microwave engineering (multiresonators) and antenna technology (cross-polarized Tx/Rx antennas), we believe to have achieved less mutual coupling effects, greater number of possible bits, and easier encoding than that reported in refs. [18, 19].

1.5 Proposed Chipless RFID System

The proposed chipless RFID system uses spectral signatures for data encoding and is fully passive, hence the tags do not need any power supply in order to operate [20]. The main application for this chipless RFID system is mainly short ranged (up to 40 cm) tagging of extremely low cost items. Hence, power limitation restrictions (transmitted EIRP maximum of −45 dBm outdoors and −55 dBm indoors) do not present a major concern for the proposed system. The principal block diagram of the proposed chipless RFID system is shown in Fig. 1.2.

As can be seen in Fig. 1.2, the chipless tag encodes data in the frequency spectrum hence having a unique ID of spectral signatures. The spectral signature is obtained by interrogating the tag by a continuous wave (CW) multi-frequency signal of uniform amplitude and phase. The tag then receives the interrogation signal and encodes the data into the frequency spectrum in both magnitude and phase. The encoded signal is then retransmitted back to the reader. This allows the reader to use two criteria for data decoding—amplitude and phase.

The chipless RFID tag consists of UWB antennas and a multiresonating circuit operating in the UWB frequency spectrum as shown in Fig. 1.3. The UWB antennas are used to receive the interrogation signal sent from the reader and transmit the signal back to the reader after performing spectral signal modulation by the multiresonator. The multiresonator is a combination of multiple filtering sections, which are used to modulate the spectrum of the interrogation signal sent by the reader. Modulation is performed in both magnitude and phase of the spectrum.

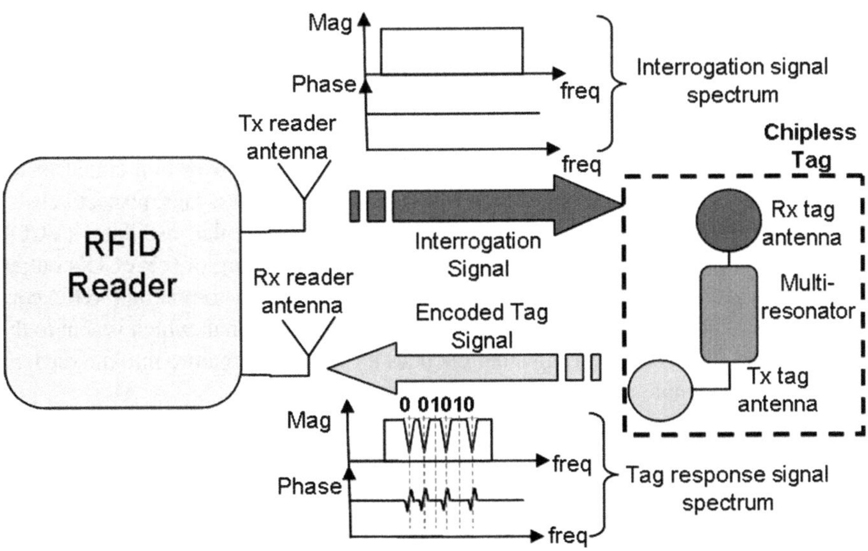

Fig. 1.2 Principal block diagram of proposed chipless RFID system

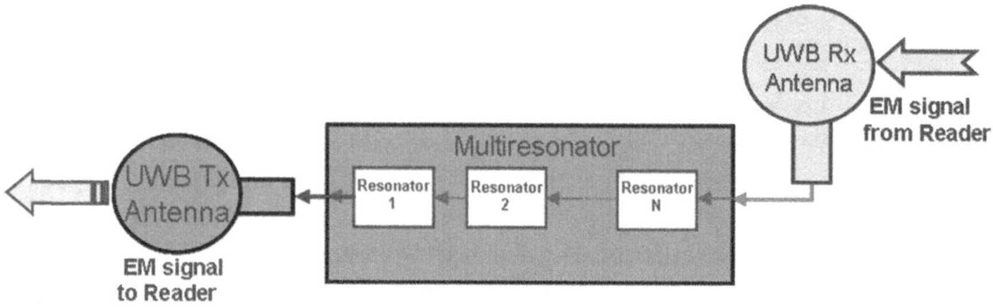

Fig. 1.3 Block diagram of proposed chipless RFID tag

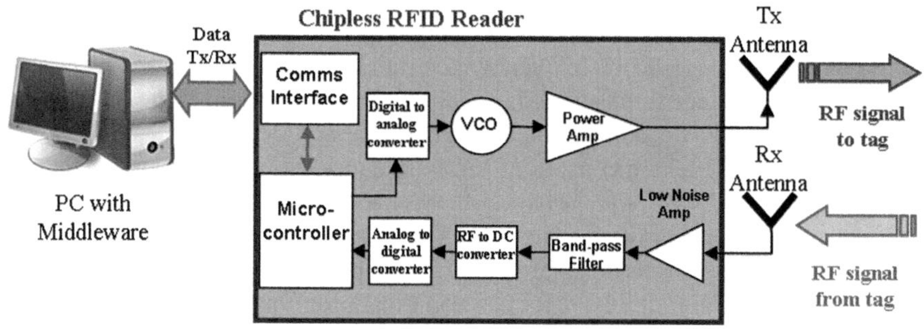

Fig. 1.4 Block diagram of proposed chipless RFID reader

The magnitude and phase are modulated in the forms of magnitude attenuations and phase jumps at the resonant frequencies of the multiresonator, respectively.

The chipless RFID reader is an electronic device, which can detect the ID of the chipless tag once it is within the reader's interrogation zone. The block diagram of the chipless RFID reader and its basic components are shown in Fig. 1.4. The RFID reader has transmitting and receiving antennas for sending the interrogation signal to the chipless tags and receiving the encoded signal from the chipless tags, respectively.

The RFID reader transmitter comprises a voltage controlled oscillator (VCO), low noise amplifier (LNA), and power amplifier (PA). Tuning of the VCO's output frequency is done by the microcontroller through the digital-to-analog (ADC) converter. The reader transmitter generates the interrogation signal, which is sent to the chipless tag. The chipless tranponder encodes its spectral signature into the reader's interrogation signal and sends the signal back to the reader.

The signal processing of the received tag signal is performed at the receiver end of the RFID reader and results in a digital signal being sent to the microprocessor of the RFID reader. The receiver comprises an LNA, a band-pass filter (BPF), a demodulating circuit which converts the RF signal to baseband, and an analog-to-digital converter. The microprocessor uses tag detection and decoding algorithms to determine the ID of the chipless tag. This data is sent to an application or software enterprise on a PC, which provides the graphical user interface between the RFID system and the user.

1.6 Book Outline

This section provides a brief description of the chapters presented in this book.

1.6.1 Chapter 1: Introduction

An introduction to RFID systems, RFID applications and the limitations of barcodes and emergence of RFID as an enabling technology is presented. Chipless RFID systems are presented as a low cost option for tagging low cost items. The proposed concept of the chipless RFID is presented followed by the outline of the chapters.

1.6.2 Chapter 2: Low Cost Chipless RFID Systems

This chapter represents the literature review, and it focuses on chipless RFID tags and readers that can be found available on the market and reported in scientific conferences and journals. A novel classification of RFID tags and readers is presented along with the system specifications of the proposed chipless RFID tag and RFID reader.

1.6.3 Chapter 3: Spiral Resonators

In this chapter, spiral resonators are studied as spectral signature encoding circuits for the novel chipless RFID tag. The theory of operation and comprehensive parametric study of two types of spiral resonators are conducted: microstrip and co-planar waveguide spiral resonator. The microstrip spiral resonator is used for tags designed on standard, printed circuit board, while the co-planar waveguide spiral resonator is used for thin flexible laminates due to its superior performance. Finally, the spiral resonators are cascaded to form a multiresonating circuit, which is used for multiple bit data encoding using the spectral signature concept.

1.6.4 Chapter 4: Ultra Wideband Antennas

Two types of UWB antennas are studied in this chapter: circular monopole and log periodic dipole array. The circular monopole antenna is used for the chipless tag since it yields UWD with an almost omnidirectional radiation pattern. The log periodic dipole antenna is used for the RFID reader since it has high gain and a directional radiation pattern. The theory and design of these antennas is presented followed by the results, which confirm the successful design and operation of the antennas.

1.6.5 Chapter 5: Chipless RFID Tag

In this chapter, the integration of the UWB monopole antennas and the multiresonating circuits is done to form the chipless RFID tag. Three types of tags are designed: 6-bit proof-of-concept tag, 35-bit tag, and 23-bit flexible RFID tag. The layout specifications and measurements of the chipless tags are presented. This chapter finalizes the design of the chipless RFID tag.

1.6.6 Chapter 6: Transceiver Design for Chipless RFID Tag Reader

The design of the RFID transceivers needed to interrogate the chipless tags previously designed is presented in this chapter. Firstly, a Gen-1 transceiver is designed, which only detects the amplitude of the tag's spectral signature. Following the Gen-1 design, the Gen-2 transceiver is designed with amplitude and phase detection capabilities. The Gen-2 transceiver is upgraded to operate with the UWB region, and hence the UWB transceiver is designed. All three transceiver circuits are tested and characterized by wired readings to the tag, and their successful operation is confirmed.

1.6.7 Chapter 7: Chipless RFID Tag-Reader System

The integration of the chipless RFID system is presented in this chapter. The chipless RFID system application is specified. The chipless RFID system comprises a chipless RFID tag and RFID reader. The RFID reader is designed by integrating the transceiver circuits and the digital control section of the reader. The chipless tag is firstly interrogated inside the anechoic chamber using a vector network analyzer and then it is tested outside a laboratory setup, which represents a reflective environment. The chipless tag is finally interrogated by the developed RFID reader to confirm successful system operation.

1.6.8 Chapter 8: Conclusions and Future Works

The conclusions focus on the important findings in this research project and highlight future works, which are mainly in the area of successful printing of tags on plastic banknotes, long range reading capabilities of the reader, and anticollision.

Chapter 2
Low Cost Chipless RFID Systems

Abstract This chapter presents a comprehensive overview and novel classification of chipless RFID tags that can be found in the market and reported in peer-reviewed conferences and journals. The problems of achieving low cost chipped RFID are presented along with the proposed solution of chipless RFID for lowering the price of the RFID tag. Two main chipless RFID tag groups are presented: time domain reflectometry-based chipless RFID and spectral signature-based chipless RFID. A comprehensive report on different chipless RFID tags with their pro's and con's is presented. Following the chipless RFID tag classification is a comprehensive classification of chipless RFID readers, which can be found on the market. A system-level view and architectural exploration of RFID readers is presented. The chapter is closed with the presentation of the proposed multiresonator-based chipless RFID tag and reader along with their requirements and specifications.

2.1 Introduction

In the preceding chapter, an introduction to RFID and chipless RFID systems was presented. The proposed chipless RFID system based on multiresonators was presented in block diagram format. The proposed chipless RFID system is a novel spectral signature design, which encodes data in both magnitude and phase.

This chapter focuses on the difficulties of achieving low cost chip RFID systems and the emergence of chipless RFID systems as a cheaper solution. The main issue of chipped RFID tags is the cost of the IC and its assembly to the printed antenna.

S. Preradovic and N.C. Karmakar, *Multiresonator-Based Chipless RFID: Barcode of the Future*, DOI 10.1007/978-1-4614-2095-8_2,

A comprehensive review of chipless RFID transponders available on the market and reported in peer-reviewed journals/conferences is presented. The novel system concept of using chipless transponders results in having new RFID readers, and their system architecture is presented in this chapter.

2.2 Difficulties of Achieving Low Cost RFID

The use of RFID instead of optical barcodes presented in Chap. 1 has not been achieved due to the greater price of the RFID tag (10 cents) compared to the price of the optical barcode (less than 0.1 cent). The arguments for not having a cheap RFID transponder are comprehensively presented by Fletcher [21]. Fletcher advocates that Application Specific Integrated Circuit (ASIC) design and testing along with the tag antenna and ASIC assembly result in a costly manufacturing process. This is why it is not possible to further lower the chipped RFID transponder price. The basic steps for manufacturing a chipped RFID transponder are shown in Fig. 2.1.

The design of silicon chips has been standardized for over 30 years, and the cost of building a silicon fabrication plant is in billions of US dollars [22, 23]. Since silicon chips are fabricated on a wafer-by-wafer basis, there is a fixed cost per wafer (around US $1,000). The cost of the wafer is independent of the IC design; hence, the cost of the RFID chip can be estimated by knowing the required silicon area for the RFID chip. Significant achievements have been made in reducing the size of the transistors allowing more transistors per wafer area [24]. Further on, decreasing the amount of transistors needed results in an even smaller silicon area, hence a lower RFID chip price. As a result, great efforts have been put in by MIT to design an RFID ASIC with less than 8,000 transistors. Although this will reduce the price of the silicon chip, its miniature size imposes limitations and further cost for handling such small chips.

The cost of dividing the wafer, handling the die and placing them onto a label is still significant even if the cost of the RFID chip is next to nothing (which it is not). The cost of handling the die increases with the use of smaller than standard chips simply because the electronics industry is not standardized for them.

Hence, it is safe to say that with highly optimized low transistor count ASICs, implemented assembly processes and extremely large quantities (over one billion) of RFID chips sold per annum a minimum cost of 5 cents is the reality for chipped RFID transponders.

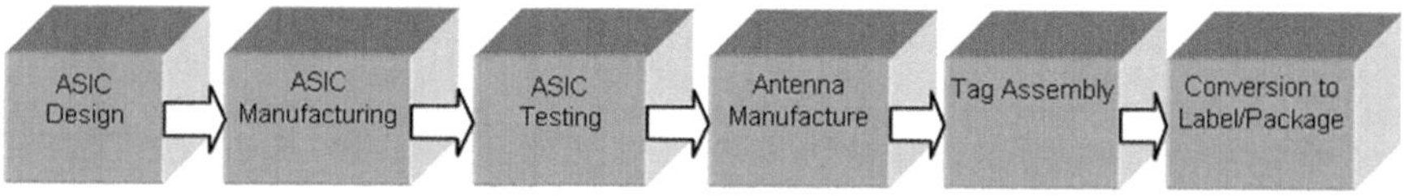

Fig. 2.1 RFID label/transponder manufacturing process

2.3 Chipless RFID Transponders: The Low Cost RFID Solution of the Future

Given the inevitable high cost of silicon chip RFID transponders (when compared to optical barcodes), efforts of designing low cost RFID transponders without the use of traditional silicon ASICs have emerged. These transponders, and therefore systems, are known as *chipless* RFID systems. Most chipless RFID systems are based on using the electromagnetic properties of materials and/or designing various conductor layouts/shapes to achieve particular electromagnetic properties/behaviour. From here on, the main focus of this book will be on chipless RFID systems.

2.3.1 Review of Chipless RFID Transponders

There have been a few reported chipless RFID tag developments in recent years (Table 2.1). However, most of them are still reported as prototypes and only a handful is considered to be commercially viable or available. The challenge that researchers face when designing chipless RFID transponders is how to perform data encoding without the presence of a chip. In response to this problem, two general types of RFID transponders can be identified: time domain reflectometry (TDR)-based and spectral (frequency) signature-based chipless RFID transponders. Figure 2.2 shows the classification of reported chipless RFID transponders.

TDR-based chipless RFID transponders are interrogated by sending a signal by the reader in the form of a pulse and listening to the echoes of the pulse sent by the tag. This way a train of pulses is created, which can be used to encode data. Various RFID transponders have been reported using TDR-based technology for data encoding. We can distinguish between non-printable and printable TDR-based transponders.

Table 2.1 Specifications for chipless RFID tag

Electrical specifications	
Frequency of operation	3.1–10.7 GHz
Tag antenna polarization	Linearly polarized
Tag antenna radiation	Preferably omni-directional
Number of bits	Greater than 20 bits
Mechanical specifications	
Tag Size	Width: 64 mm maximum ; Length: 120 mm
Printability	Fully printable, no semiconductor
Operations	Printed on thin plastic/paper objects
Weight	Less than 5g
Visibility	Preferably transparent
Operating temperature	–20° to 80°C
Commercial	
Cost	Less than 1 cent per tag.

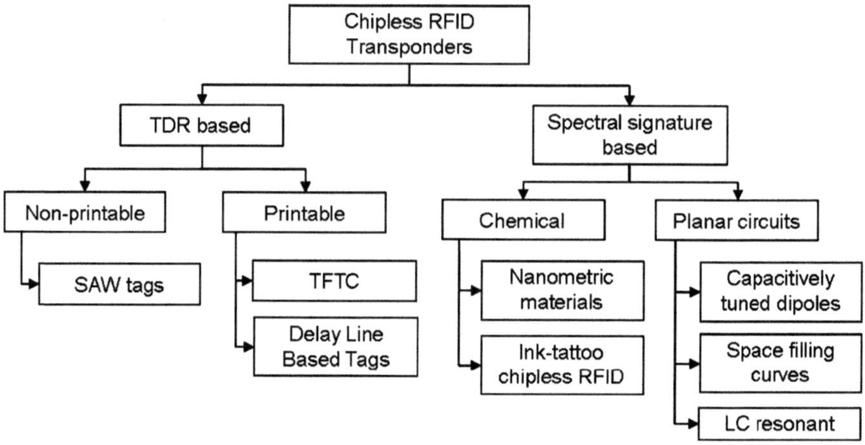

Fig. 2.2 Classification of chipless RFID transponders

The example of a *non-printable* TDR-based chipless RFID transponder is the surface acoustic wave (SAW) tag, which is also the commercially most successful developed by RFSAW Inc. [25]. *SAW tags* are excited by a chirped Gaussian pulse sent by the reader centred around 2.45 GHz [26–30]. The interrogation pulse is converted to a SAW using an interdigital transducer (IDT). The SAW propagates across the piezoelectric crystal and is reflected by a number of reflectors, which creates a train of pulses with phase shifts [31–38]. The train of pulses is converted back to an EM wave using the IDT and detected at the reader end, where the tag's ID is decoded [39–48].

Printable TDR-based chipless transponders can be found either as Thin-Film-Transistor Circuits (TFTC) or microstrip-based transponders with discontinuities. *TFTC transponders* are printed at high speed and on low cost plastic film [49]. TFTC tags offer advantages over active and passive chip-based transponders due to their small size and low power consumption. They require more power than other chipless tags, but offer more functionality. However, low cost manufacturing processes for TFTC tags have not been developed yet. Another issue is the low electron mobility, which limits the frequency of operation up to several MHz.

Delay-line based chipless RFID tags operate by introducing a microstrip discontinuity after a section of delay line as reported in refs. [50–52]. The transponder is excited by a short pulse (1 ns) EM signal. The interrogation pulse is received by the transponder and reflected at various points along the microstrip line creating multiple echoes of the interrogation pulse. The time delay between the echoes is determined by the length of the delay line between the discontinuities. This type of tag is a replica of the SAW tag using microstrip technology, which makes it printable. Although initial trials and experiments of this chipless technology have been reported, only 4 bits of data have been successfully encoded, which shows limited potential of this technology.

Spectral signature-based chipless transponders encoded data into the spectrum using resonant structures. Each data bit is usually associated with the presence or absence of a resonant peak at a predetermined frequency in the spectrum. So far, five types of spectral signature-based tags have been reported and all five are considered to be fully printable. We can distinguish two types of spectral signature tags based on the nature of the tag: chemical tags and planar circuit tags.

Chemical transponders are designed from a deposition of resonating fibres or special electronic ink. Two companies from Israel use *nanometric materials* to design chipless tags. These tags consist of tiny particles of chemicals, which exhibit varying degrees of magnetism and when electromagnetic waves impinge on them they resonate with distinct frequencies, which are picked up by the reader [53]. They are very cheap and can easily be used inside banknotes and important documents for anti-counterfeiting and authentication. CrossID, an Israeli paper company, claims to have such 70 distinct chemicals, which would thus provide unique identification in the order of 2^{70} (over 10^{21}) when resonated and detected suitably [54]. Tapemark also claims to have "nanometric" resonant fibres, which are 5 μm in diameter and 1 mm in length [55]. These tags are potentially low cost and can work on low grade paper and plastic package material. Unfortunately, they only operate at frequencies up to a few kHz, although this gives them very good tolerances to metal and water.

Ink-tattoo chipless tags use electronic ink patterns embedded into or printed onto the surface of the object being tagged. Developed by Somark Innovations [56], the electronic ink is deposited in a unique barcode pattern, which is different for every item. The system operates by interrogating the ink-tattoo tag by a high frequency microwave signal (>10 GHz) and is reflected by areas of the tattoo, which have ink creating a unique pattern which can be detected by the reader. The reading range is claimed to be up to 1.2 m (4 feet) [57, 58]. In the case of animal ID, the ink is placed in a one-time-use disposable cartridge. For non-animal applications, the ink can be printed on plastic/paper or within the material. Based on the limited resources available for this technology (still in experimental phase), the author has assumed that it is spectral signature based.

Planar circuit chipless RFID transponders are designed using standard planar microstrip/co-planar waveguide/stripline resonant structure, such as antennas, filters and fractals. They are printed on thick, thin and flexible laminates and polymer substrates. *Capacitively tuned dipoles* were first reported by Jalaly [59]. The chipless tag consists of a number of dipole antennas, which resonate at different frequencies. When the tag is interrogated by a frequency sweep signal, the reader looks for magnitude dips in the spectrum as a result of the dipoles. Each dipole has a 1:1 correspondence to a data bit. Issues regarding this technology would be: tag size (lower frequency longer dipole—half wavelength) and mutual coupling effects between dipole elements.

Space-filling curves used as spectral signature encoding RFID tags were first reported by McVay [60]. The tags are designed as Peano and Hilbert curves with resonances centred around 900 MHz. The tags represent a frequency selective surface (FSS), which is manipulated with the use of space-filling curves (such as Hilbert curve, etc.). The transponder was successfully interrogated in an anechoic chamber. Only 3 bits of data have been reported so far. However, the tag requires significant layout modifications in order to encode data.

LC resonant chipless tags comprise a simple coil, which is resonant at a particular frequency. These transponders are considered 1-bit RFID transponders. The operating principle is based on the magnetic coupling between the reader antenna and the LC resonant tag. The reader constantly performs a frequency sweep searching for transponders. Whenever the swept frequency corresponds to the transponder's resonant frequency, the transponder will start to oscillate producing a voltage dip across the reader's antenna ports. The advantage of these tags is their price and simple structure (single resonant coil), but they are very restricted in operating range, information storage (1 bit), operating bandwidth and multiple-tag collision. These transponders are mainly used for electronic article surveillance (EAS) in many supermarkets and retail stores [61].

In the following section, the general RFID reader architecture and modern RFID reader review will be presented.

2.4 Modern RFID Readers

RFID readers are devices that perform the interrogation of RFID transponders. In a chipless RFID system, the RFID reader detects the tag by using signal processing demodulation techniques to extract data from the transponder's signal. A chipless tag cannot generate a signal without the reader sending an interrogation signal to the transponder. Therefore, the reader and transponders are in a master–slave relationship, where the reader acts as a master and the transponders as slaves. Nevertheless, RFID readers themselves are in a slave position as well. A software application, also called middleware, processes data from the RFID reader, acts as the master unit and sends commands to the reader [62] as shown in Fig. 2.3.

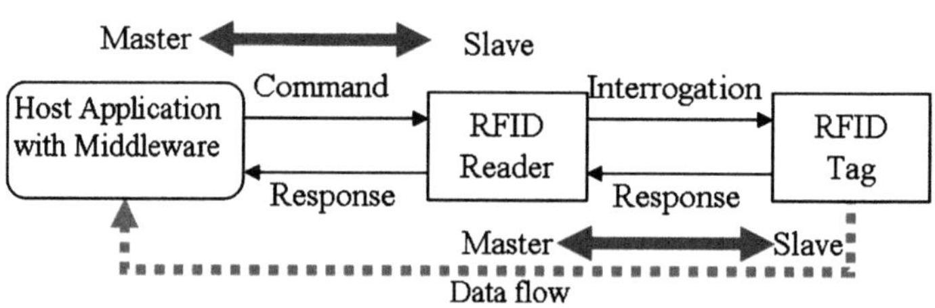

Fig. 2.3 Master–slave principle between the application software and reader, and the reader and transponders

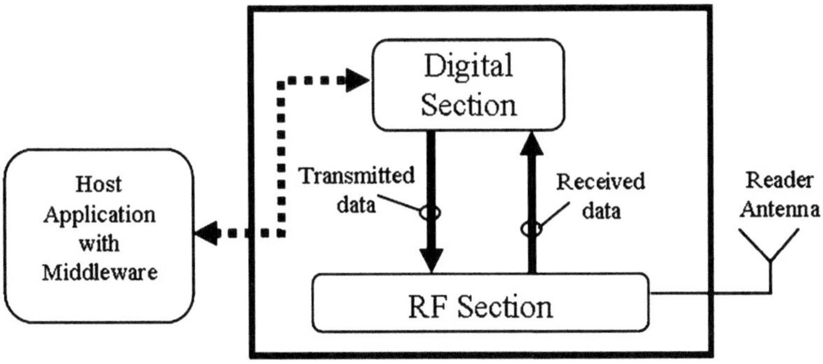

Fig. 2.4 Block diagram of a typical RFID reader

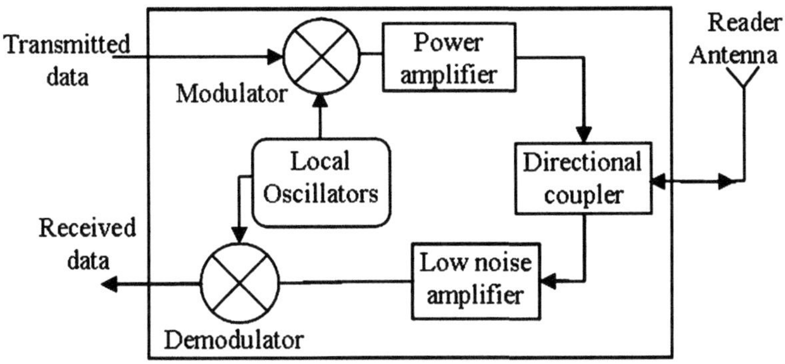

Fig. 2.5 Block diagram of the RF section of an RFID reader

2.4.1 RFID Reader Architecture

An RFID reader consists of three main parts as shown in Fig. 2.4. These main three components are:

1. Digital/control section.
2. RF section.
3. Antenna.

The digital section of the RFID reader performs digital signal processing over the received data from the RFID transponder. This section usually consists of a microprocessor, a memory block, a few analogue-to-digital converters (ADCs) and a communication block for the software application.

The reader's RF section is used for RF signal transmission and reception and consists of two separate signal paths to correspond with the two directional data flows as shown in Fig. 2.5.

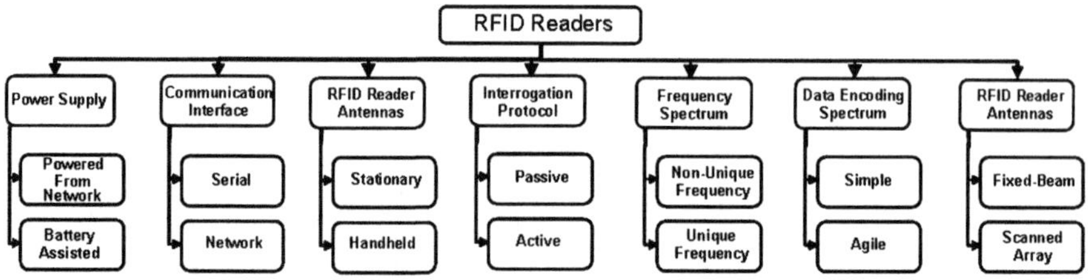

Fig. 2.6 Classification of RFID readers available in the market and open literature

The local oscillator generates the RF carrier signal, a modulator modulates the signal, the modulated signal is amplified by the power amplifier, and the amplified signal is transmitted through the antenna. A directional coupler separates the system's transmitted signal and the received weak back-scattered signal from the tag [63]. The weak back-scattered signal is amplified using low noise amplifiers (LNA) before the signal is decoded in the demodulator. Different demodulation techniques are used when decoding the data received from the transponder. Most RF sections are protected from EM interference using metal cages.

Depending on the RFID system's applications, the RFID reader can be designed in different ways, where the antenna's resonating frequency, gain, directivity and radiation pattern can vary. Adaptive antennas act as spatial filters and a promising technique for implementing this spatial diversity into RFID readers [64]. The antenna reported in ref. [63] is a 5-element rectangular patch antenna array with an intelligent beam forming network at 2.45 GHz. A number of different reader antennas have been developed during the years based on microstrip patch antennas [65–67].

Following is a detailed discussion on various RFID reader systems available in the market and reported in open literatures.

2.4.2 *Review of RFID Readers*

Figure 2.6 shows the classifications of RFID readers available in open literature and commercial markets. The classification is done after an analysis and synthesis of a comprehensive literature review on RFID readers. The classification is based on the power supply, communication interface, mobility, tag interrogation, frequency response and the supporting protocols of the reader.

The classification of the RFID reader based on their power supply brings forth two types of readers: readers supplied from the power network and battery-powered (BP) readers.

Readers supplied by the *power network* generally use a power cord connected to an appropriate external electrical outlet. Most readers that use this type of power supply are fixed stationary readers and their operating power supply ranges from 5 to 12 V [68], but there are examples of readers that operate at voltage levels as high as 24 V [69].

Battery-powered readers are light in weight and portable. The battery is mainly used to power up the motherboard of the reader. Most BP readers are hand-held, but there are stationary readers that are battery assisted as well. BP readers use from 5 V up to 12 V batteries for their power supply [70, 71].

Based on their *communication interface*, readers can be classified as serial and network.

Serial readers use a serial communication link to communicate with their host computers or software applications. The reader is physically connected to a host computer using the RS-232 [72], RS-485, I2C or USB serial connection [73–75].

Network readers are connected to the host computer via a wired or wireless network. These types of readers behave like a standard network device. Today's RFID readers support multiple network protocols, such as Ethernet, TCP/IP, UDP/IP, HTTP, LAN, WLAN and others [76, 77]. This allows easier tracking, maintenance, data rate and results in a smaller number of hosts for installing a large number of readers in comparison with serial readers.

The next classification of RFID readers can be made on their *mobility*. Hence, we distinguish two types of readers: stationary and hand-held RFID readers.

Stationary RFID readers are also known as fixed readers. This term comes from the reader's ability to be mounted on walls, portals, doors or other objects, where they can perform effective transponder readings and are not meant to be moved or carried. Fixed RFID readers are mainly used for wireless data capture in supply chain management, asset tracking and product control [78], but can also be found in personnel identification and authentication for restricted access areas [79–81] as well.

Hand-held RFID readers are mobile readers that can be carried and operated by users as hand-held units. Hand-held readers have built-in antennas and usually do not have connectors for additional antennas. They are BP and are light weight (from 82 g up to 700 g). They have shorter reading ranges than fixed readers [82, 83] and are mainly used for tracking live stock, locating items in stores and in stock, etc.

Another classification of RFID readers can be made upon the reader's *interrogation protocols* in terms of being passive and active.

Passive readers are limited to only "listening" and do not perform additional tag interrogations. When interrogating the tags, the reader sends CW signal as a power source for the RFID transponder, which becomes activated. Upon activation, the RFID transponder transmits its unique ID back to the reader [84].

Active readers are true interrogators, which interrogate and listen to tags. Active readers perform data transmission towards the tags, which is implemented, in most cases, as a modulation of the carrier signal. Therefore, transponders must have a demodulating circuitry enabling them to decode the reader's command. These readers can perform both listening and calling out to the tags and can even achieve

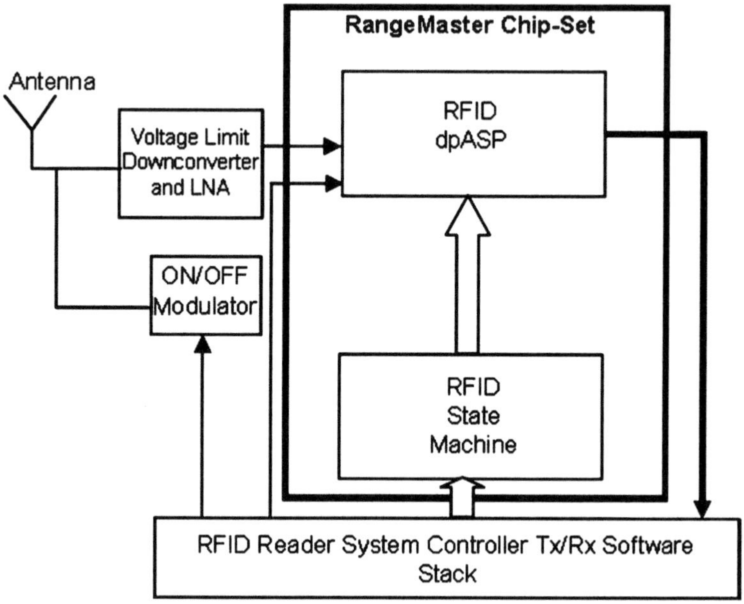

Fig. 2.7 System-level overview of the RangeMaster-embedded RFID reader

successful area location of the transponder based on the amplitudes of the transponders respond to the reader's interrogation [85].

We can classify readers based on the *transponder frequency responses* that they listen to as unique frequency response and non-unique frequency response-based readers.

Unique frequency response-based readers operate at a unique (or short bandwidth <80 MHz) frequency range and use this frequency for both data transmission and reception. The vast majority of RFID readers that can be found on the market today are unique frequency response-based readers.

None-unique frequency response-based readers operate using one frequency for sending a command or just provide a carrier signal at a certain frequency and listen for an integer multiple of its carrier frequency, generally in the form of a second harmonic, or a frequency-divided signal as the transponder's response [86]. Two RF frequencies used for communication by the reader to the RFID system allow fast and reliable full-duplex communication, but this system needs a more complex RF front end for both the reader and transponder module [87]. Figure 2.7 shows a multi-frequency RFID system.

We can distinguish between two types of RFID readers based upon their ability to communicate with transponders in regards to *data-encoding protocols*: simple RFID readers and agile RFID readers.

Simple RFID readers use a unique protocol for communication and data transmission between transponders in the reader's interrogation zone [88, 89]. When a transponder that supports the reader's interrogation protocol is set in the interrogation area of the reader, the tag is automatically recognized and detected. When a transponder that operates using a different protocol is put into the interrogation area, no data transmission will occur because of the unfamiliarity of the reader's interrogation protocol to the transponder.

Agile RFID readers can operate and perform interrogations and data transmission with transponders using multiple protocols. The most commonly used protocols for data transmission between transponders and readers are EPC Gen1 [90], EPC Gen2 [91], ISO 18000 [92], TIRIS Bus Protocol, etc. The majority of RFID readers that can be found in the market are designed for multiple protocols and multi-tag readings [93].

We can distinguish between two types of RFID readers based on their *antennas*: fixed beam RFID readers and scanned array RFID readers.

Fixed beam antennas are characterized with a unique and fixed beam radiation pattern [94]. The use of several fixed beam antennas is used as well and can be commonly found in Alien Technology readers. The advantage of using such antennas is that they are easy to install and do not need any logic to control their radiation patterns. The disadvantage of these antennas is that they pick up multipath signals alongside with the transponders signal, which can result in errors during interrogation.

Scanned Array RFID readers use smart antenna systems in order to reduce the number of transponders within their main lobe radiation zone, thus reduce reading errors and collisions among tags. This technique exploits spatial diversity among tag's locations. The direct beam also reduces the effects of multipathing [95]. This new approach to RFID antenna technology is being incorporated by a few RFID manufacturers, such as Omron Corporation, Japan [95], RFID Inc. [96] and RFSAW, USA [89].

2.4.3 Towards Universal Reader Design

RFID reader designers and manufacturers have gone a step further in the design of independent reader modules [97, 98]. The design of embedded RFID readers has been introduced to the world of RFID in 2005 [99]. In June 2005, Anadigm Inc. announced the birth of the industry's first RFID-embedded reader that can be customized to read different RFID tag types, with different modulation schemes, frequencies and data transmission protocols [100]. The "universal" reader is named RangeMaster. The RangeMaster's system-level block diagram is shown in Fig. 2.7. It comprises a Field Programmable Analogue Array (FPAA) in conjunction with an RFID State Machine, enabling RFID system engineers to develop a universal RFID reader supporting multiple protocols and frequencies for future fixed, mobile and hand-held reader designs [101]. The advantage of embedded RFID readers, such as RangeMaster, when compared to standard readers is that they allow standardization around a single circuit board, simplification and improvement of product development, manufacturing time and cost.

2.5 Chipless RFID System Specifications

Several RFID system design requirements are discussed in this paragraph. These requirements are largely determined by the application for which the RFID system will be deployed.

1. *Cost.* The cost of an RFID system is largely dependent on the cost of the transponder. The chipless RFID tag needs to be extremely cheap—below 1 cent would be preferable in order to be affordable when tagging low cost paper-/ plastic-based items. This places restrictions on both transponder design and the choice of materials for its construction. Typical conductors that can be used are copper, aluminium and conductive ink. Typical dielectrics are polyester and PCB substrates.
2. *Size.* The size of the transponder is dependent on both frequency of operation and the size of the tagged item. The size of the transponder should be from several centimetres to approximately a decimetre.
3. *Frequency band.* The frequency band of operation is an important aspect of the proposed system because it directly determines the amount of bits, which can be encoded. The proposal for a UWB system is the chosen option. The UWB frequency spectrum varies from country to country (generally from 3.1 to 10.7 GHz, USA). UWB systems are restricted by the amount of EIRP, which is in the noise level (below −40 dBm outdoors and below −50 dBm indoors).
4. *Read range.* Minimum required reading range is specified based on the reader sensitivity, which is entirely due to the fact that the chipless transponder does not need power supply for operation. Limitations in reading range will be introduced due to the low EIRP and orientation.
5. *Application with mobility.* The chipless RFID system is intended for conveyor belt applications at speeds of 10 m/s. The Doppler shift, in this case, at 10.7 GHz (worse case scenario for UWB system) is less than 400 Hz and does not effect the operation of the chipless transponders. Although the transponder spends less time in the interrogation zone of the reader, since the transponder and reader do not communicate using handshaking or synchronization, this aspect does not present a major issue.
6. *Reliability.* RFID transponders should be reliable devices that can sustain variations in heat, humidity, stress and processes, such as printing, label insertion and lamination. Conductive ink has proven to be extremely robust and when printed on flexible substrates, it maintains its robustness which is a necessity for applications, such as envelope or note-bill tagging.
7. *Security.* The chipless RFID system can provide an extra layer of security against counterfeiting when using transparent conductive inks. A transponder may be printed without visible notice. Another advantage would be the fact that it would be impossible to read multiple tags stacked together (a stack of tagged note bills cannot be read accurately due to the mutual coupling and the inability to differentiate between tags or determining the number of tags).

The following sections will present the proposed chipless RFID transponder and RFID reader circuits.

2.6 Proposed Chipless RFID Tag

The review of available and reported chipless RFID transponders has shown the lack of an operational fully printable multibit chipless RFID transponder. This section presents the proposed novel chipless RFID transponder based on multiresonators. The main components of the transponder are the transmitting (Tx) and receiving (Rx) antennas and multiresonating circuit. Block diagram of the integrated chipless RFID transponder with basic components is shown in Fig. 2.8.

The proposed chipless RFID transponder consists of a vertically polarized UWB disc-loaded monopole receiving (Rx) tag antenna, a multiresonating circuit and a horizontally polarized UWB disc-loaded monopole transmitting (Tx) tag antenna.

When the interrogation signal reaches the transponder, it is received using the receiving monopole antenna and propagates further on towards the multiresonating circuit. The multiresonating circuit encodes data bits using cascaded spiral resonators, which introduce attenuations and phase jumps at particular frequencies of the spectrum. After passing through the multiresonating circuit, the signal contains the unique spectral signature of the tag and is transmitted back to the transmitter using the transmitting monopole tag antenna. The receiving and transmitting tag antennas are cross-polarized in order to minimize interference between the interrogation signal and the re-transmitted encoded signal containing the spectral signature.

The main differences between our spectral signature-based transponder and the ones reported in the previous section are that we encode data in both amplitude and phase, the transponder operates in the UWB region and that the tag responses are not based on radar cross-section (RCS) backscattering, but on retransmission of the cross-polarized interrogation signal with the encoded unique spectral ID.

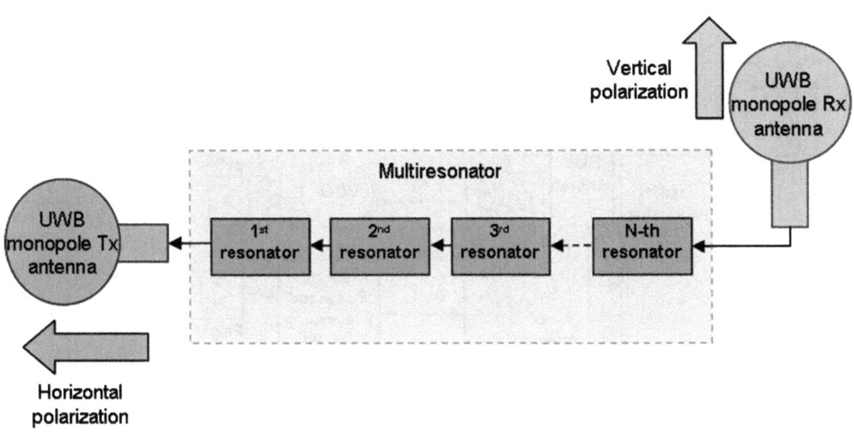

Fig. 2.8 Chipless RFID transponder circuit block diagram

2.7 Proposed Chipless RFID Reader

In this book, three types of RFID readers will be presented: first Generation (Gen1) chipless RFID reader, second Generation (Gen2) chipless RFID reader and the UWB RFID reader (Table 2.2).

The Gen1 and Gen2 RFID readers are designed as the first and second generation RFID chipless tag reader prototypes operating between 1.9 and 2.5 GHz. The two main parts of the RFID reader are: the Digital section and the RF section. A block diagram of the RFID reader is illustrated in Fig. 2.9.

Table 2.2 Specifications for chipless RFID tag reader

Electrical specifications	
Frequency of operation	3.1–10.7 GHz
Transmitting power	15 dBm
Number of reader antennas	At least two antennas—transmitting and receiving
Reader antenna polarization	Linearly polarized
Reader antenna gain	Above 5 dBi
Communication interface	Serial bus interface (RS-232)
Mechanical specifications	
Connector	SMA/DB-9
Operations	Mountable over a conveyor belt
Size	Portable—within 200 mm by 200 mm
Weight	Less than 5 kg
Environmental	Suitable for work in industrial environment
Operating temperature	–20° to 50°C
Commercial	
Cost	Less than AU $2000 (a guide only)

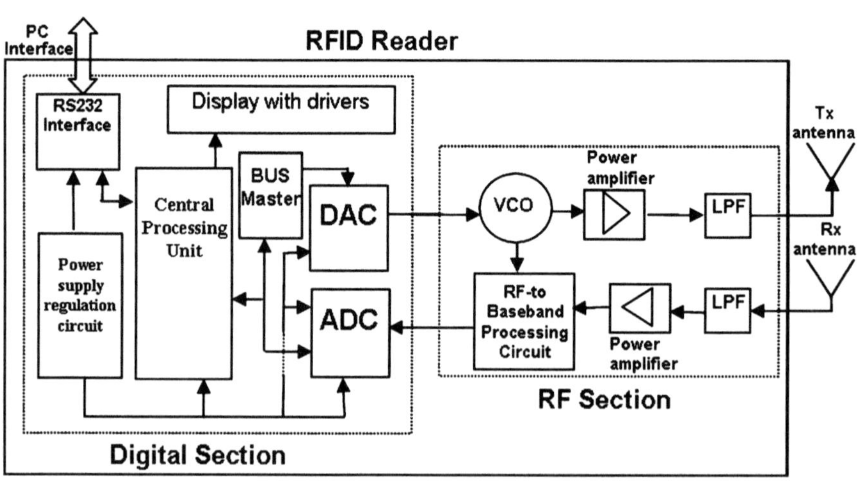

Fig. 2.9 Block diagram of Gen1 and Gen 2 chipless RFID reader

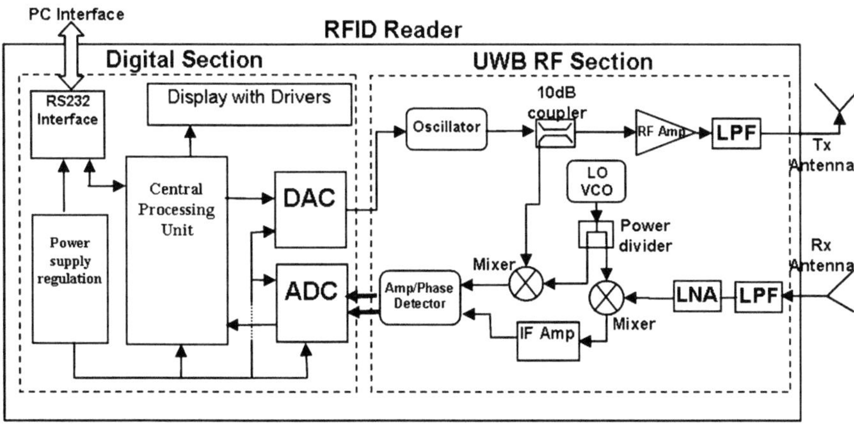

Fig. 2.10 Block diagram of proposed UWB chipless RFID reader (7–9 GHz)

The central processing unit (CPU) sends data to the RF section using a digital-to-analogue (DAC) converter. The analogue data is the tuning voltage for the voltage-controlled oscillator (VCO), which generates the RF signal for interrogating the chipless tag. The received signal from the tag is amplified and filtered before being sent to the RF-to-Baseband (RTB) circuit, where it is converted to a DC analogue value and sent to the digital section. The analogue signal is then converted to a digital signal using the ADC and sent to the CPU for tag ID decoding. The tag ID is displayed on a 7-segment LED display section and/or sent to the host computer.

The digital section is designed around an 8-bit Atmel AT89C52 microcontroller [110], which performs the major signal processing and data-decoding algorithms. The RF section consists of two RF paths: transmitter and receiver. The transmitting circuit is comprised of a VCO, which generates the RF interrogation signal and is used as a reference signal for the receiver. The interrogation signal is thus amplified and filtered and transmitted by a wideband directive reader antenna. The signal is received by a cross-polarized receiver antenna and processed (filtered and amplified) before being sent to the RTB circuit. The RTB circuit differs between the Gen1 and Gen2 RFID reader. The Gen1 reader uses magnitude detection, hence a diode rectifier is used as the RTB circuit. The Gen2 reader utilizes a Gain/Phase detector AD8302 [111] as the RTB. The AD8302 compares the received RF signal from the tag with the reference signal from the VCO and yields DC equivalent values of magnitude and phase difference between the two RF signals. The two DC values are multiplexed and sent to the digital section of the reader for further processing (digitizing) and decoding.

The UWB RFID reader is upgraded to work in the UWB region by introducing a wide-band oscillator and down-converting mixers as shown in Fig. 2.10. The mixers are used to down-convert the UWB signals (from 5 to 9 GHz to below 2.5 GHz), where the gain/phase detector AD8302 operates. The received tag signal and reference signal are compared in amplitude and phase by the AD8302. The amplitude and phase difference are given as separate DC values by the AD8302. The 2 DC values are multiplexed and then sent to the digital section, where they are digitized and processed further on for tag ID detection.

With the design proposals of the chipless RFID reader and transponder, we can conclude this chipless RFID system specification. The following chapters will focus on the tag and reader design and experimental results.

2.8 Conclusions and Motivation

In this chapter, we have presented novel classifications of RFID transponders and RFID readers. The classification was based on open literature and published research articles. As shown, there are various types of RFID systems. The majority of RFID transponders are chipped based, and hence require power supply in order to operate. This puts limitations in terms of cost and implementation.

A fully printable chipless RFID transponder which can be used on paper-/plastic-based items has not been yet developed. Most chipless transponders found in the review are either in early prototyping stages or have been there for quite a while (5 years or so). This leaves an open area for research where a novel and operating chipless RFID system can be developed. The proposed chipless RFID transponder based on multiresonators was presented. The transponder is comprised of two cross-polarized UWB monopole antennas, a multiresonating circuit.

The advantages of fully printable chipless transponders are the fact that they have the potential to be extremely low cost (less than 1 cent), robust (no chip—no mechanical damage), secure (using transparent conductive ink), zero-power consumption and zero maintenance.

The next step was the proposal of a circuit for interrogating the chipless transponders—a chipless RFID reader. The hardware and software development of a fully operational reader enables the proposal of the chipless RFID system to transform into a potential application, which could be commercialized or optimized for tagging particular items (mainly, conveyor belt applications where items are scanned one at a time).

The successful development of this chipless RFID system will be the most cost-effective tagging solution for low cost paper/plastic items and will provide an extra layer of security against counterfeiting, passport and sensitive documents control, enable cheap and effective tracking of currency and open the door towards the era of ultra-low cost printable chipless RFID.

As mentioned in Sect. 2.6, the transponder requires narrowband resonators in order and UWB antennas in order to encode as many bits possible and provide enough frequency spectrum for fitting all of the bits. The resonances are obtained by utilizing fully printable planar microwave resonant structure—resonators. UWB tag antennas are designed as linear-polarized UWB monopole antennas, which are fully planar. Hence, the following chapters will focus on the design of the tag multiresonator and UWB antenna.

Chapter 3
Spiral Resonators

Abstract The spiral resonator is one of the main components of the chipless RFID tag, which is used to encode data into the tag's spectral signature. In this chapter, we present a model of the spiral resonator based on lumped components and transmission lines. A parametric study of the spiral resonator's different layout parameters is conducted in order to design the most optimal spiral resonator. Spiral resonators are developed in microstrip and coplanar waveguide technology. Coplanar waveguide technology provides superior performance on thinner laminates. A novel data-encoding technique using spiral shorting is presented. Finally, a multiresonating circuit comprised of cascaded spiral resonators used to encode data into the spectral signature is designed and optimized.

3.1 Introduction

Having set the objectives, goals and system specifications in the preceding chapter, the following chapters present the component level design of the chipless tag, system level design of the RFID readers followed by the system integration and field trials. This chapter presents the detailed investigation of the spiral resonator, which is the fundamental component of the chipless RFID tag. According to the proposed chipless RFID tag presented in Chap. 2, the tag encodes data by using a multiresonator. The multiresonator consists of cascaded microwave spiral resonators coupled to a microstrip transmission line. The microwave spiral resonators must be fully planar, exhibit narrow bandwidth meaning high Q factor and compact in size for use in the printable chipless tag.

In this chapter, a microwave resonant structure which can meet the requirements presented in Chap. 2 is investigated. The most suitable candidate that can meet all the desirable features is the spiral resonator [102]. When compared to other planar circuits, such as defected ground structures (DGS) [103–112], spiral resonators have the advantage of having 5–50 times narrower 3 dB bandwidth. It can also be

S. Preradovic and N.C. Karmakar, *Multiresonator-Based Chipless RFID: Barcode of the Future*, DOI 10.1007/978-1-4614-2095-8_3,

modified easily for data encoding that will be shown later on in this chapter. Various spiral resonators coupled to the microstrip lines can be found in open literature. Some are etched in the ground plane [113], while others are etched inside the microstrip line [114] or gap coupled to the microstrip line [115].

The conventional microstrip spiral resonator gap coupled to the microstrip line [114] is used due to its narrow bandwidth (20 MHz), reasonably high attenuation at its resonant frequencies (>7 dB) and single-sided layout. Understanding the theoretical models of the spiral is the foundation for designing high Q spiral resonators. The theoretical model of the spiral resonator is discussed in detail later in this chapter.

In this chapter, the theoretical model and operation of the spiral resonator are presented. The distributed component model and coupled line model are presented. A parametric study of the spiral resonator is performed. Many useful design curves are derived from the study. Following the parametric study, the microstrip spiral resonators are designed on 90-μm thin flexible laminate to emulate the tagged polymer banknote. However, the attenuation of the spiral resonator on the thin laminate drops drastically (700%) due to the fact that the energy concentration inside the thin laminate is very high and the skin depth may be comparable to the thickness of the copper clad. This problem has been properly addressed and solved by transferring the design to a coplanar waveguide (CPW) technology. A novel data-encoding technique is introduced that can ease mass production of tags with unique ID's. Finally, the loading effects of the multiple resonators on a transmission line and their mutual coupling have been investigated thoroughly followed by the conclusions.

3.2 Theoretical Modelling of Spiral Resonator

Figure 3.1 shows the layout of a conventional spiral resonator. The microstrip line and the spiral resonator are on the same plane (top layer) and are separated from the continuous metallic ground plane (bottom layer) by a dielectric layer. From Fig. 3.1, it is clear that the spiral resonator is gap coupled to the 50 Ω microstrip line. At its resonant frequency, the spiral resonator creates a stop-band effect.

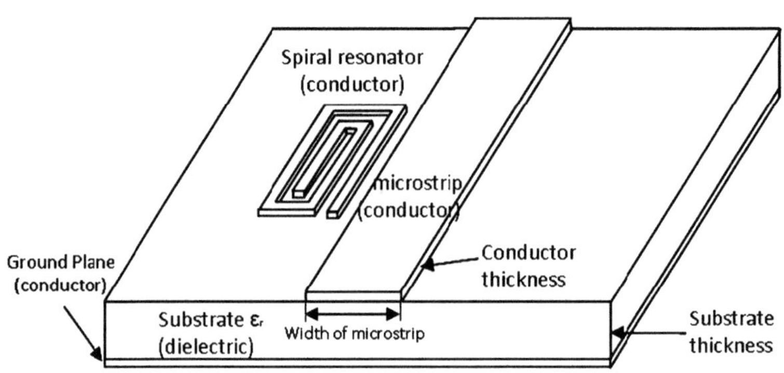

Fig. 3.1 Layout of spiral resonator placed next to a microstrip line

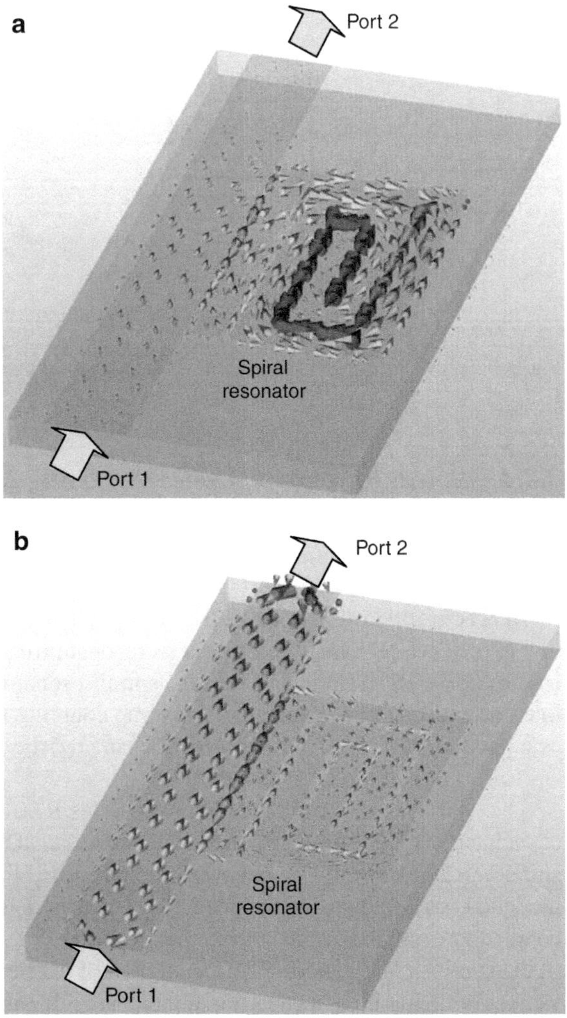

Fig. 3.2 CST surface current distribution of spiral resonator at (**a**) resonant frequency of 2 GHz and (**b**) non-resonant frequency of 2.1 GHz

The surface current distribution simulation is used in order to understand how the stop-band effect is created at the spiral's resonant frequency. Figure 3.2 shows (a) the peak surface current distribution of a spiral resonator at its resonant frequency (2 GHz) and (b) at a non-resonant frequency (2.1 GHz). The simulation was performed using CST Microwave Studio 2008. From Fig. 3.2a, it is clear that the surface current distribution is greater around the spiral at its resonant frequency. The spiral resonator creates a low impedance path to ground at its resonant frequency and absorbs the majority of the current propagating from Port 1 to Port 2 of the microstrip line resulting in a stop-band effect. At its resonant frequency, the spiral resonator couples almost none of the surface current propagating between Port 1 and Port 2 as seen in Fig. 3.2b.

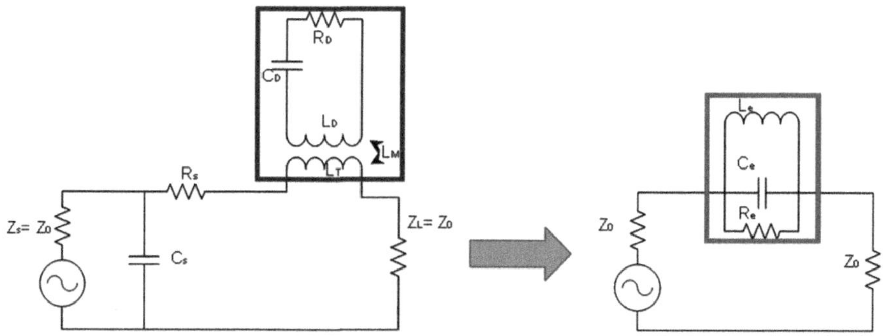

Fig. 3.3 Equivalent circuit model spiral resonator coupled to microstrip line [116]

3.2.1 *Spiral Resonator Modelling Using Distributed Components*

The spiral resonator coupled to a microstrip line can be modelled using distributed capacitance and inductance as a series RLC circuit coupled to a microstrip line [116]. The spiral resonator on its own is modelled as a series RLC circuit due to the low impedance path that it creates at its resonant frequency, which is the characteristic of series RLC circuits. When the spiral resonator is coupled to the microstrip line, the entire circuit (microstrip line + gap coupling + spiral resonator) is modelled as a parallel RLC due to its stop-band characteristic as reported in ref. [116]. The equivalent circuit model is shown in Fig. 3.3.

The spiral resonator is modelled as a series RLC circuit, where the distributed capacitance is CD, the distributed inductance is LD and the resistive loss of the spiral is RD. The coupling between the microstrip line and the spiral resonator is modelled with mutual inductance LM, which is determined by the length of the coupled line and the distance from the 50 Ω line.

Jiang et al. [117] reported the calculation of distributed capacitance and resonant frequency of spiral resonators. First, the charge distributions on the spiral resonator are found by a simplified model and the moment method. Knowing the charge distribution, it is possible to calculate the distributed capacitance of one spiral turn C_1 as:

$$C_1 = \frac{1}{2}q = \pi H \sum_{i=1}^{N_1} r_i \rho_i. \tag{3.1}$$

The spiral turn surface is subdivided into N_1 sub-annuli, the mean radius of the spiral is r, charge density is ρ and charge quantity is q. The total equivalent distributed capacitance of the spiral resonator

$$C_\mathrm{D} = \frac{C_1(R_\mathrm{i} + R_\mathrm{o})}{2r}, \tag{3.2}$$

where R_i is the inner radius of the spiral and R_o is the outer radius of the spiral.

Hejazi et al. [118] reported the calculation of distributed inductance of spiral resonators. The reported method for calculating the distributed inductance of spiral resonators calculates the inductance of individual turns and the mutual inductance between turns of the spiral resonator. Applying this method, the total distributed inductance L_D of any spiral shape has the general form of:

$$L_D = \sum_{i=1}^{n} L_{oi} + 2\times\left[\sum M^{+} - \sum M^{-} + \sum_{j=1}^{j} M_{j,j+1} \pm \sum_{k=1}^{k} M_{k,k+2}\right], \tag{3.3}$$

where L_{oi} are the self-inductances of the turns, M^{+} is the mutual inductance between turns when the current is in concurrent directions, M^{-} is the mutual inductance between turns when the current is in opposite directions, $M_{j,j+1}$ is the mutual inductance between adjacent segments meeting at a point and the nearest segments not meeting at a point $M_{k,k+2}$, $j=1, 2, \ldots, n-1$ and $k=1, 2, \ldots, n-2$ and n is the maximum number of turns in the spiral resonator.

The resonant frequency f_r of the spiral resonator coupled to a microstrip line is reported in ref. [115], [119] as:

$$f_r = \frac{1}{2\pi\sqrt{L_e C_e}}, \tag{3.4}$$

where L_e and C_e are the total equivalent inductance and capacitance of the band stop filter, respectively.

3.2.2 Spiral Resonator Modelling Using Coupled Lines

The previous section focused on spiral resonator modelling using distributed lumped components. This method does provide an accurate RLC circuit model of the spiral resonator. However, it does not significantly aid the designer when designing the layout of the spiral resonator. A coupled line theory model of the spiral resonator is proposed in Figs. 3.4 and 3.5. A transmission line model of the spiral resonator models the frequency characteristics of the resonator based on the layout of the resonator and not the equivalent lumped circuitry.

Figure 3.4 shows the sections of the spiral resonator, which are used to model it in ADS Schematic Simulator. There are a total of 22 sections of the spiral resonator. The sections of the spiral resonator are represented by schematic microstrip components as shown in Table 3.1. However, it is important to notice that ADS Schematic does not support coupled microstrip bends, hence this type of circuit could not be used in the transmission line model of the spiral resonator. Instead, the lack of a coupled microstrip bend circuits was compensated with asymmetrically coupled line circuits and 90° bends. The transmission line model still gave very accurate simulation results when compared to the simulation results obtained using ADS Momentum. These results will be shown in the following sections.

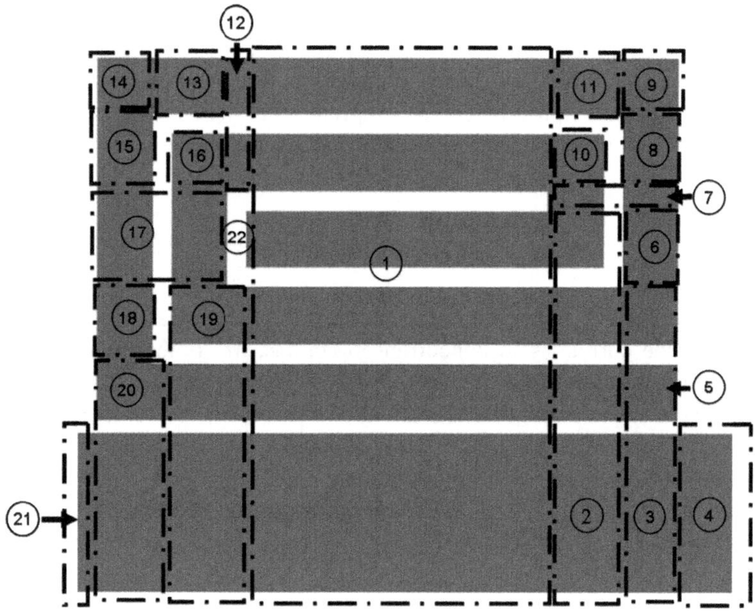

Fig. 3.4 Spiral resonator circuit divided into sections

Figure 3.6 shows the conventional design phase of a spiral resonator. The design phase starts with the choice of a laminate on which the spiral resonator is to be designed on. Further, an approximation of the spiral resonator's layout parameters/ dimensions is done based on the quarter wavelength of the desired resonant frequency. The full wave solver is then used to calculate the S-parameters vs. frequency behaviour of the frequency resonator. If the results are not satisfactory, the spiral resonator dimensions are varied, and the full wave analysis is performed again until satisfactory results are achieved. Obviously, this type of design method is time consuming and purely iterative.

The coupled line model (Fig. 3.5) enables faster and easier design of the spiral resonator. The optimization time is much shorter due to the fact that transmission line theory-based simulators take less time to simulate than that for the full wave 3D or Method of Moments (MoM) solvers do. The transmission line model has been designed in ADS 2008 Schematic Editor. It is important to note that the model has neglected the influence of coupling between the microstrip bends due to the non-existence of such microstrip components in the ADS Schematic Editor.

The following section focuses on the parametric study of the spiral resonator and the comparison between the performances of the spiral resonators obtained using a commercial full wave solver (ADS Momentum) and the proposed coupled line model of the spiral resonator. This study facilitates not only understanding of the operation of the spiral resonator, but also deriving many useful design curves. These design curves are used to tune the resonator into the optimal design.

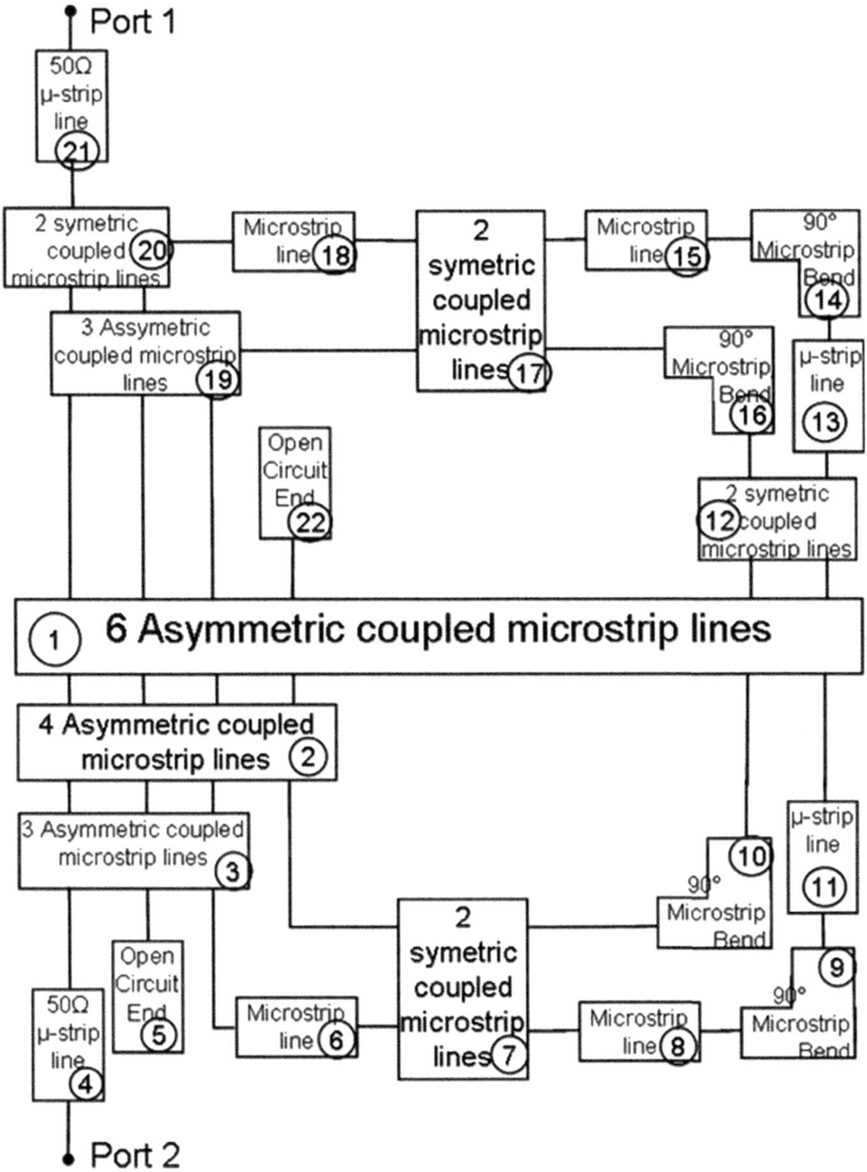

Fig. 3.5 Transmission line model of spiral resonator based on coupled lines

Table 3.1 ADS schematic components used in the transmission line model of spiral resonator

ADS schematic components	Section no.
Six asymmetric coupled lines	1
Four asymmetric coupled lines	2
Three asymmetric coupled lines	3, 19
Two symmetric coupled lines	7, 12, 17, 20
Microstrip line	4, 6, 8, 11, 13, 15, 18, 21
90° Microstrip bend	9, 10, 14, 16
Open circuit	5, 22

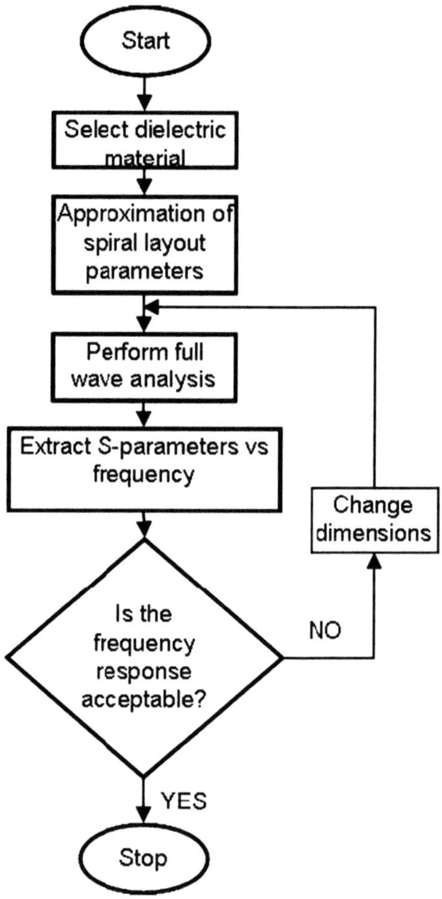

Fig. 3.6 Conventional steps for spiral resonator design

3.3 Parametric Study of Microstrip Spiral Resonator on PCB

As can be seen from Fig. 3.7, the layout of a conventional microstrip spiral resonator is defined by the following parameters: W_{feed} is the width of the microstrip line, D_{gap} is the separation between the spiral and microstrip line, $W_{spiralcond}$ is the width of the

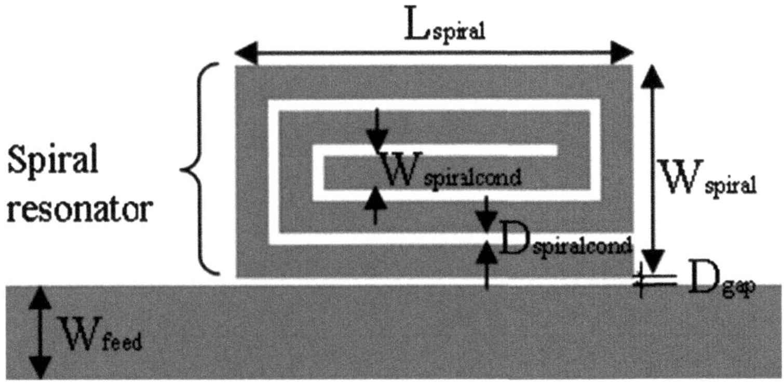

Fig. 3.7 Layout of spiral resonator with defined layout parameters

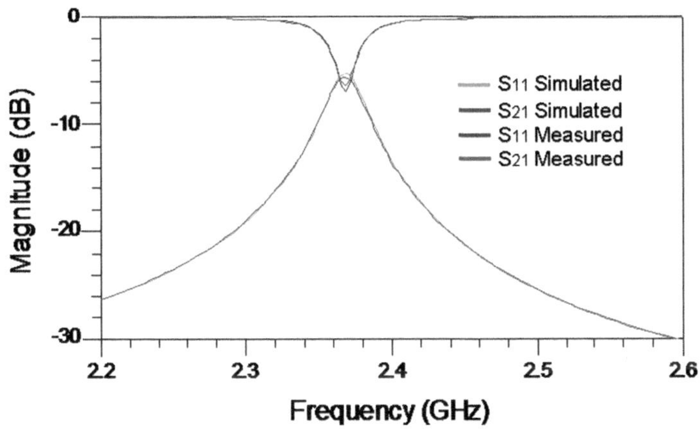

Fig. 3.8 Simulated and measured frequency response of spiral resonator designed on Taconic TLX-0 ($W_{\text{feed}} = 2.26$mm, $W_{\text{spiralcond}} = 0.8$mm, $D_{\text{spiralcond}} = 0.3$mm, $D_{\text{gap}} = 0.2$mm=0.2 mm, $L_{\text{spiral}} = 8.64$mm, $W_{\text{spiral}} = 5.2$mm=5.2 mm, εr=2.45, h=0.787 mm, tanδ=0.0019)

spiral conductor, $D_{\text{spiralcond}}$ is the separation between the spiral conductors, L_{spiral} and W_{spiral} are the length and width of the spiral resonator, respectively. The frequency response of a spiral resonator (bellow) is shown in Fig. 3.8.

Figure 3.8 shows good agreement between measured and simulated results obtained by ADS Momentum 2008. In the simulation, all boundaries are considered to be open space, and the ports are set to have 50 Ω impedance. The S21 curve of the spiral resonator shows resonance at 2.37 GHz with 7 dB attenuation. The corresponding S11 curve has a magnitude of −5.2 dB. The S21 3 dB bandwidth is 20 MHz. The spiral resonator is designed on Taconic TLX-0.

The first parameter that is varied in the parametric study is the number of turns N in the spiral resonator. The other design parameters are kept constant, while the number of turns is varied from 1 to 4. The width of the spiral resonator conductor is

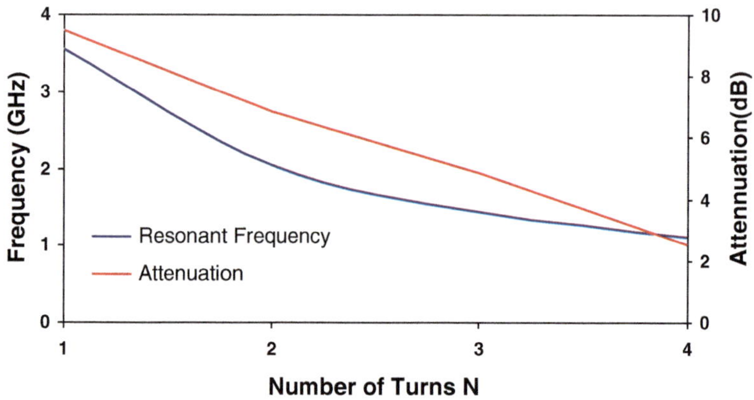

Fig. 3.9 Variation of resonant frequency and attenuation at resonant frequency with number of turns N (other parameters are as for Fig. 3.8)

0.8 mm, separation between spiral conductors is 0.3 mm and gap separation between the spiral and microstrip line is 0.2 mm, while the spiral length is kept at 10 mm. The parametric study is shown in Fig. 3.9. From Fig. 3.9, it is clear that the insertion loss/attenuation of the spiral resonator decreases with the number of turns. Although the greatest insertion loss at the resonant frequency is for $N=1$ another important parameter must be taken into account and that is the length of the spiral resonator L_{spiral}. We can see that for the same spiral length with greater number of turns the resonant frequency decreases. This is as expected due to the increase of the trace length of the spiral resonator. Therefore, we see that for a smaller number of turns the insertion loss is greater, but the length of the spiral has to be greater too. In order to have compact size spiral resonator a spiral with two turns is chosen since its length and attenuation provide the optimal solution.

After choosing the number of turns for the spiral resonator, the transmission line model presented in Sect. 3.2.2 is benchmarked against an existing MoM solver—ADS 2008 Momentum. A parametric study of the spiral resonator is performed using the proposed transmission line model based on coupled line theory and ADS 2008 Momentum. The spiral layout parameters that are varied are: (1) the length of the spiral (L_{spiral}), (2) separation between spiral conductors ($D_{\text{spiralcond}}$), (3) spiral conductor width ($W_{\text{spiralcond}}$) and (4) distance of the spiral from the microstrip line (D_{gap}). The dielectric which is used is Taconic TLX-0 with the following properties: $\varepsilon r=2.45$, $h=0.787$ mm, $\tan\delta=0.0019$.

Figures 3.10 and 3.11 show the variation of the resonant frequency and insertion loss with length of the spiral resonator, respectively. The width of the spiral resonator conductor is 0.2 mm, separation between spiral conductors is 0.3 mm and gap separation between the spiral and microstrip line is 0.2 mm. As expected, the resonant frequency of the spiral resonator increases as the length of the spiral resonator decreases and vice versa. The insertion loss of the spiral resonator is mainly constant from 2 to 4 GHz. As it can be seen, the transmission line model in ADS Schematic Editor based on coupled lines corresponds quite nicely to the simulation results with Momentum.

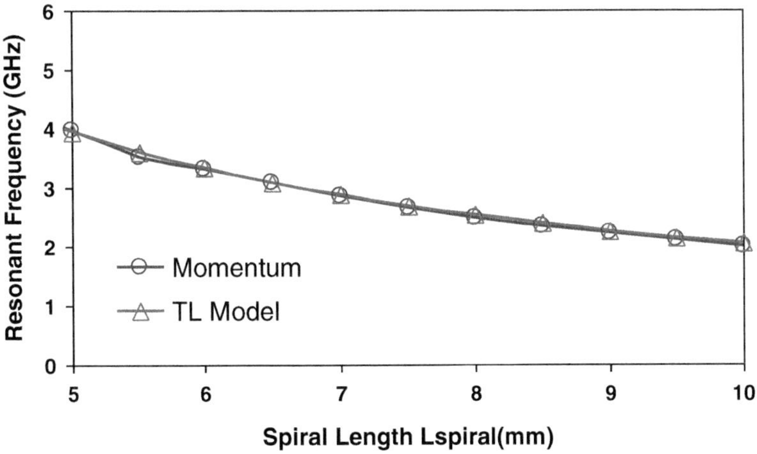

Fig. 3.10 Variation of attenuation vs. spiral length L_{spiral} (other parameters are as for Fig. 3.8)

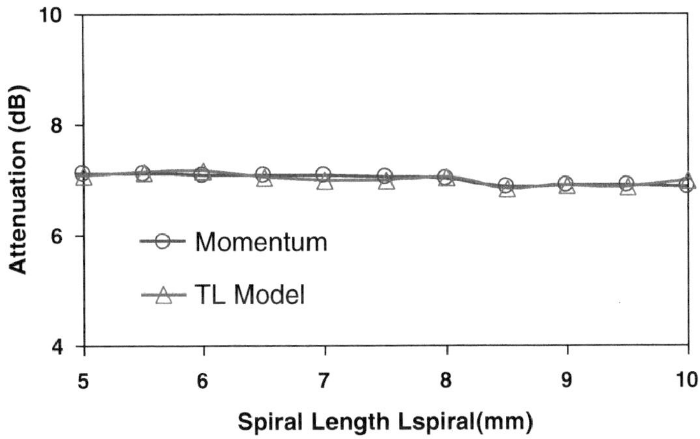

Fig. 3.11 Variation of attenuation with separation between spiral turns $D_{spiralcond}$ (other parameters are as for Fig. 3.8)

Figures 3.12 and 3.13 show the variation of the resonant frequency and attenuation with the separation between spiral conductors (spiral turns), respectively. The width of the spiral resonator conductor is 0.8 mm, the length of the spiral resonator is 10 mm and gap separation between the spiral and microstrip line is 0.2 mm. By decreasing the gap between spiral conductors, we decrease the overall dimensions of the resonator and its trace length, which results in the increase of the resonant frequency. From Figs. 3.12 and 3.13, it is clear that the transmission line model corresponds well to the simulation results in Momentum.

Figures 3.14 and 3.15 show the variation of the resonant frequency and attenuation with the width of the spiral conductor, respectively. The length of the spiral conductor is 10 mm, the separation between spiral conductors is 0.3 mm and gap

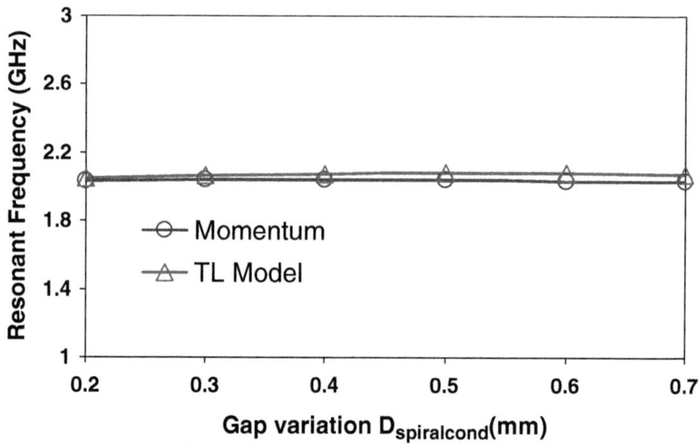

Fig. 3.12 Variation of resonant frequency vs. separation between spiral turns $D_{spiralcond}$ (other parameters are as for Fig. 3.8)

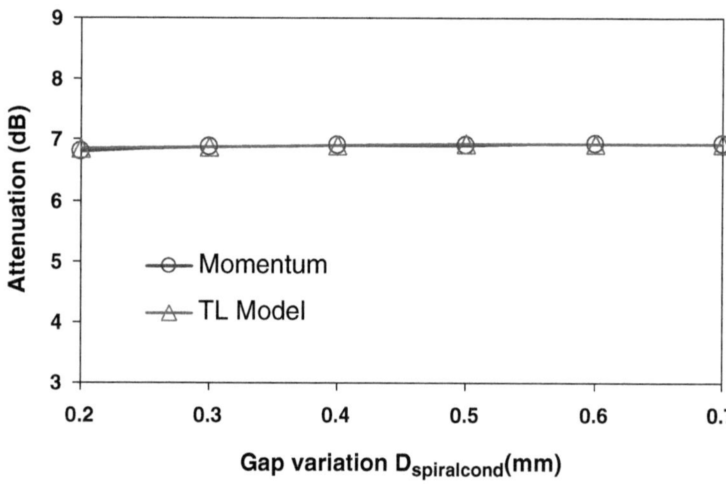

Fig. 3.13 Variation of attenuation with separation between spiral turns $D_{spiralcond}$ (other parameters are as for Fig. 3.8)

separation between the spiral and microstrip line is 0.2 mm. From Fig. 3.14, we can see that the width of the spiral conductor does not influence the shift in resonant frequency greatly. However, the spiral resonator's attenuation at the resonant frequency is greatly dependent on the spiral conductor width. The optimal spiral conductor width is 0.8 mm. Based on Figs. 3.14 and 3.15, we can conclude that the simulation results obtained with Momentum and the transmission line model designed in ADS Schematic Editor correspond nicely.

Figures 3.16 and 3.17 show the variation of the resonant frequency and attenuation with the variation of D_{gap}. The parametric study of spiral resonators has shown

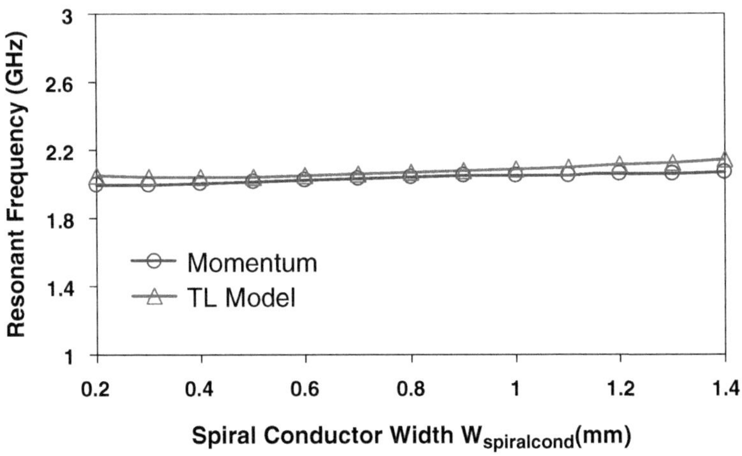

Fig. 3.14 Variation of resonant frequency with spiral conductor width $W_{spiralcond}$ (other parameters are as for Fig. 3.8)

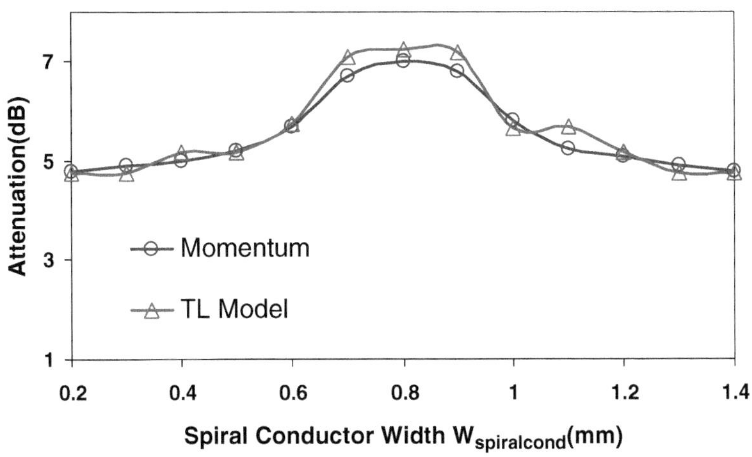

Fig. 3.15 Variation of attenuation with spiral conductor width $W_{spiralcond}$ (other parameters are as for Fig. 3.8)

that the attenuation greatly depends on $W_{spiralcond}$ and D_{gap}. The parameter D_{gap} is the gap between the spiral and the microstrip line. As expected, the attenuation of the spiral resonator is greater for smaller values of D_{gap} due to the stronger coupling between the spiral and the 50 Ω microstrip line. The variation of the resonant frequency is minimal for the different D_{gap}. Based on Figs. 3.16 and 3.17, we can conclude that the simulation results obtained with Momentum and the transmission line model designed in ADS Schematic Editor correspond nicely.

The following section presents the problems with microstrip spiral resonators printed on thin flexible laminates.

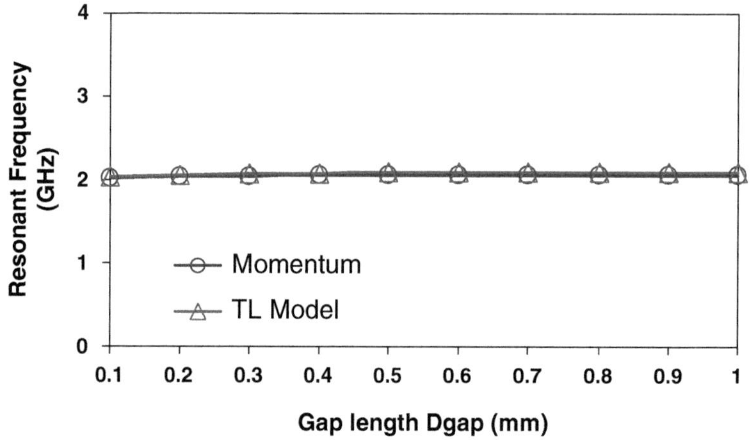

Fig. 3.16 Variation of resonant frequency with D_{gap} (other parameters are as for Fig. 3.8)

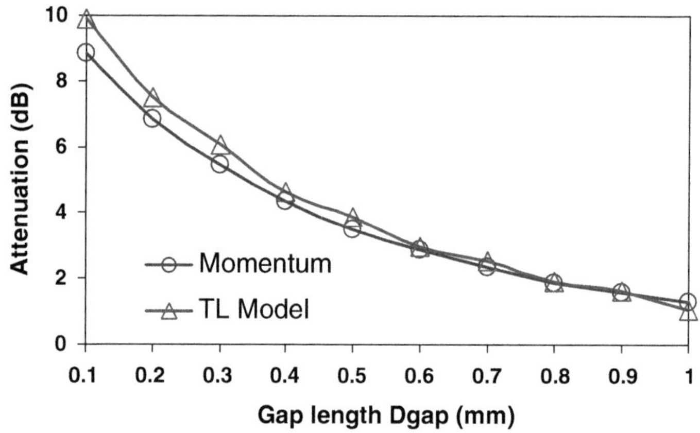

Fig. 3.17 Variation of the attenuation with D_{gap} (other parameters are as for Fig. 3.8)

3.4 Problems of Migration to Thin Flexible Substrates

The design of the spiral resonator on thick PCB was presented comprehensively in Sect. 3.3. However, it is necessary to investigate the properties of microstrip spiral resonators on thin flexible laminates. This is necessary due to the fact that the chipless tag will ultimately be printed on thin and flexible laminates, such as paper and/or polymer/plastic. The main problems of designing spiral resonators on flexible substrates are:

- Resonators exhibit low Q factor due to thin substrate.
- Microstrip lines and tracks become thinner on thinner substrates.

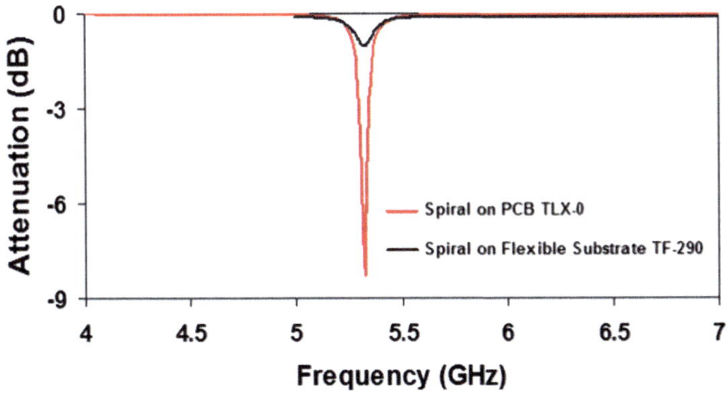

Fig. 3.18 Variation of the microstrip spiral insertion loss when printed on PCB TLX-0 (εr = 2.45, h = 0.787 mm, tanδ = 0.0019) and flexible TF-290 (εr = 2.9, h = 0.09 mm, tanδ = 0.0028)

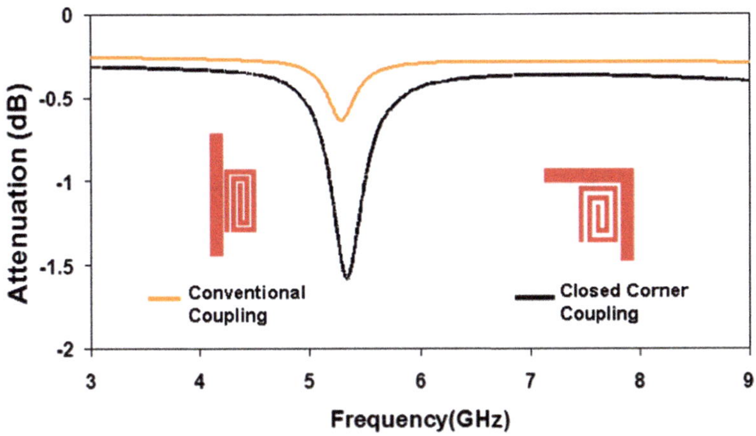

Fig. 3.19 Simulated attenuation using two different coupling methods

These two aspects make the design unrealistic for the application of polymer banknotes.

The problem with using microstrip resonators on thin flexible laminates is that the spiral resonators' quality (Q) factor drops radically as shown in Fig. 3.18, rendering them unusable for tags on flexible substrates. From Fig. 3.18, it is clear that the insertion loss drops over 700% when the spiral resonator is designed on TF-290 (90 μm thickness) in respect to a spiral resonator designed on TLX-0 (787 μm thickness).

In order to increase the attenuation of the individual spiral resonators, we used a new type of coupling to the microstrip line—corner coupling. By using corner coupling, we were able to increase the attenuation at a design frequency, as shown in Fig. 3.19. The simulations were performed using ADS Momentum 2008, hence corner coupling between the spiral resonator and the microstrip line was accounted for.

Fig. 3.20 Layout of four same spirals designed in order to increase attenuation at their resonant frequency

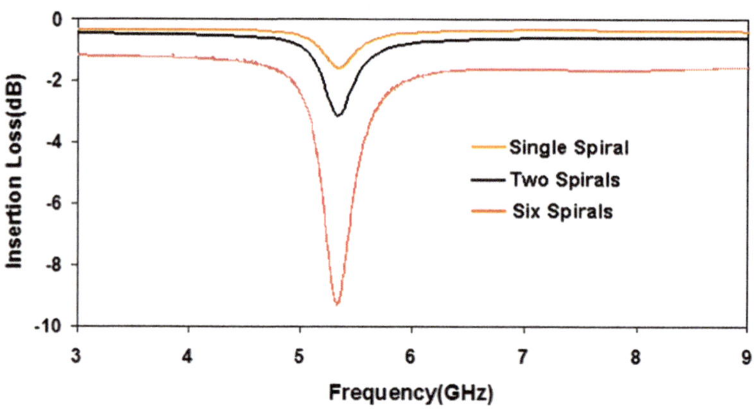

Fig. 3.21 Simulated insertion losses using one, two and six spiral resonators in order to increase attenuation

Although attenuation increased from 0.5 to 1.5 dB, it still did not reach the desired −7 dB. A tandem of spiral resonators resonating at the same frequency further increased the attenuation of the resonator as shown in Fig. 3.20. By using the same spiral resonator coupled to the microstrip line more than once (four spirals in Fig. 3.20), we increased the attenuation at a level of 2.5 dB. Figure 3.21 shows the simulated insertion losses of one, two and six spirals operating at the same resonant frequency. The disadvantage of this solution is the layout size and inefficiency (same spiral repetition), increased 3 dB attenuation bandwidth (>500 MHz). The fact that it is extremely difficult to achieve exact replicas of the same spirals, hence frequency shifts occur resulting in lesser attenuation and even larger 3 dB bandwidth.

Since neither solution seems to provide satisfactory results, the design of the spiral resonators was shifted from microstrip technology to CPW technology. CPW technology was proposed first by Dr. Sushim Mukul Roy in Monash University RFID research group, and it offers spiral resonators which are single sided, have satisfactory Q factors and have compact size. The following section will focus on designing the spiral resonator on thin flexible TF-290 laminate using CPW technology.

3.5 CPW Spiral Resonator for Chipless Tag on Flexible Substrate

The CPW was first proposed by Wen in 1969 [120]. We proposed the novel waveguide as a dielectric substrate coated with a single layer of copper on the top. The isometric view of a CPW is shown in Fig. 3.22. It consists of a conductor centre strip with conductive ground plane sheets on both sides of the strip. The impedance of the strip is determined by the width of the strip, the gap, the permittivity and thickness of the dielectric. Practically, this means that we can have multiple widths of the CPW strip on the same dielectric, which would have, for example, a 50 Ω impedance at the cost of modified the gap between the strip and ground planes. This property of CPW makes it a very flexible transmission line technology to work with.

In the previous section, it was shown that when using thin flexible laminates the microstrip spiral resonators experienced a great drop in Q factor, which resulted in smaller attenuation at their resonant frequency. Since it is crucial for the tag to be designed on thin laminates/substrates and still have a distinct dip in the spectrum magnitude this type of frequency response of the resonator is unsatisfactory.

In order to avoid the increase in metallic loss in the microstrip line due to the decrease in dielectric thickness, CPW technology is used for designing high Q spiral resonators. CPW technology uses spiral shapes etched out in the stripline to create stop bands. CPW technology overcomes the problem of low Q factor as it was obtained in microstrip technology. The layout of a spiral resonator designed on 90-μm thin Taconic TF-290 laminate is shown in Fig. 3.23.

The CPW strip line, the ground plane and the spiral resonator are on the same plane (top layer). The strip line is separated from the continuous metallic ground planes by a gap. At its resonant frequency, the spiral resonator creates a stop band as can be seen in Fig. 3.24. The 2-port S-parameters of the CPW spiral resonator

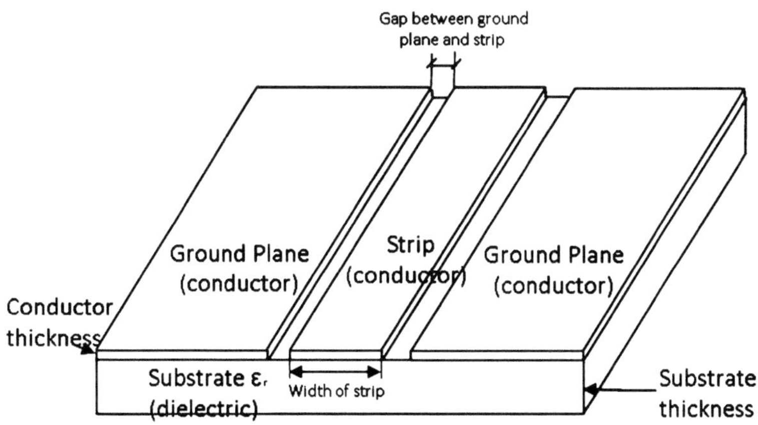

Fig. 3.22 A coplanar waveguide (CPW)

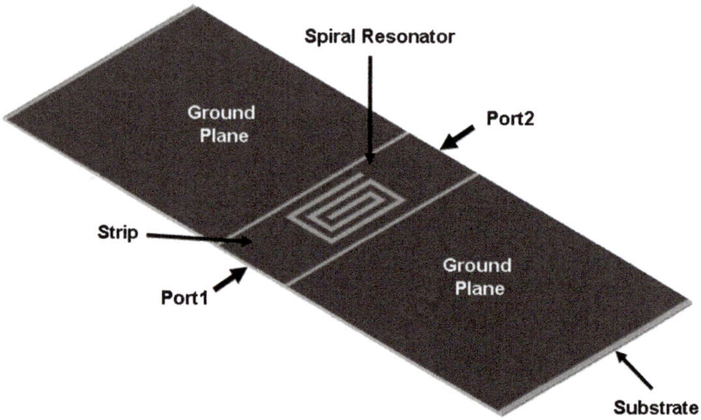

Fig. 3.23 Layout of spiral resonator etched out in a coplanar waveguide (CPW) strip line

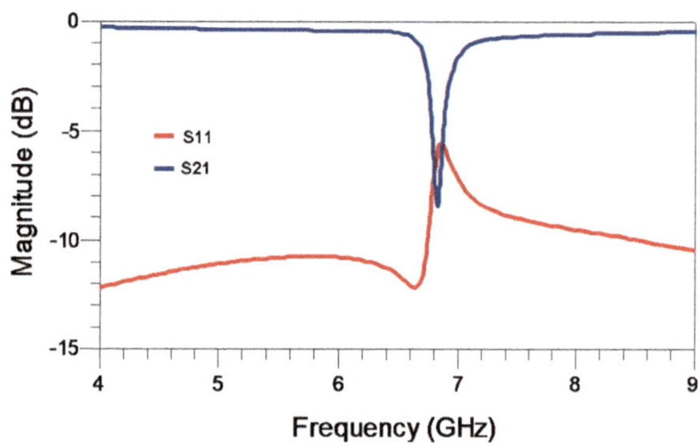

Fig. 3.24 Simulated frequency response of spiral resonator etched out in a coplanar waveguide (CPW) strip line on TF-290 (εr = 2.9, h = 90 μm, tanδ = 0.0028)

were obtained from ADS Momentum 2008. The CPW spiral resonator is designed on Taconic TF-290 (εr = 2.9, h = 90 μm, tanδ = 0.0028).

When printing the spirals using CPW technology, the stop-band attenuation is comparable with PCB results on flexible substrates as shown in Fig. 3.24 (>7 dB). This is the main reason for the migration of the tag to CPW technology. Each spiral resonator introduces a different stop-band resonance. By varying the dimensions of the spiral resonator, we can vary the resonance. Figure 3.25 shows the variation of the spiral's resonant frequency and attenuation with the spiral's resonator length L_{spiral} as obtained from ADS 2008 Momentum simulation.

The advantages of the CPW spiral resonator in comparison with the microstrip spiral resonator are higher attenuation at resonant frequency and single-sided layout, while the disadvantages would be in terms of compact layout, since CPW spiral resonators can be cascaded only by placing them in series contrary microstrip spiral resonators can be cascaded on both sides of the microstrip line.

The following section will focus on cascading spiral resonators in microstrip and CPW technology and the design of multiresonators on PCB (thick laminate) and thin flexible laminate.

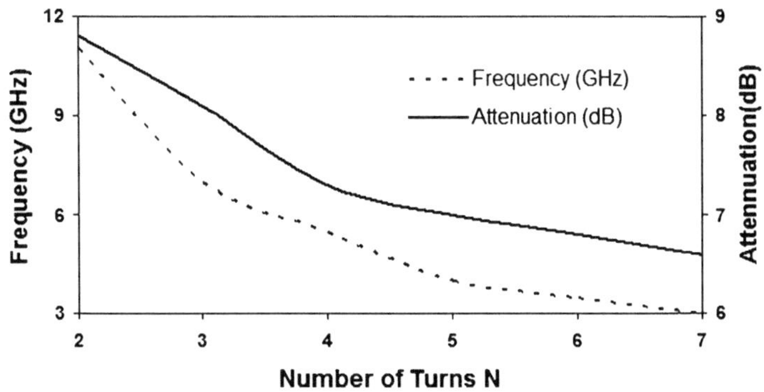

Fig. 3.25 Coplanar waveguide (CPW) spiral resonant frequency and attenuation vs. spiral length L_{spiral}

3.6 The Multiresonator: Cascaded Spiral Resonators

The theory, design, parametric study and optimization of the spiral resonators for PCB and thin flexible laminates were comprehensively investigated in the preceding chapter. However, the main task is to design a multiresonating circuit, which can encode a unique spectral signature of the tag into the interrogation signal sent by the RFID reader. The multiresonating circuit is designed by cascading spiral resonators next to the microstrip line (microstrip technology for PCB) or etching them out in the CPW strip line (CPW technology for thin laminates) with different lengths so that a multiple resonances occur. Based on this idea, the design of a microstrip multiresonator operating on PCB Taconic TLX-0 and the CPW multiresonator designed on a thin flexible laminate TF-290 are presented in this section.

3.6.1 Multiresonator on PCB Using Microstrip Technology

Figure 3.26 shows the top view of a 6-bit multiresonator layout generated in ADS Momentum 2008. Figure 3.27 shows the photograph of the fabricated 6-bit multiresonator on Taconic TLX-0 substrate. The 6-bit multiresonator consists of six spiral resonators cascaded next to a 50 Ω microstrip line. The multiresonator provides six distinguishable resonances between 2 and 2.5 GHz. Each resonance is separated by approximately 100 MHz from each other. In order to design the spirals at different frequencies, the length of each spiral has been varied so that the spiral's resonant frequency is fine-tuned. The parametric study of the spiral resonators was comprehensively presented in Sect. 3.3. The spiral resonators were placed at a separation of 3 mm from each other, had a copper conductor width of 0.8 mm and were placed 0.2 mm from the 50 Ω microstrip line. The width of the 50 Ω microstrip line on Taconic TLX-0 was calculated to be 2.26 mm.

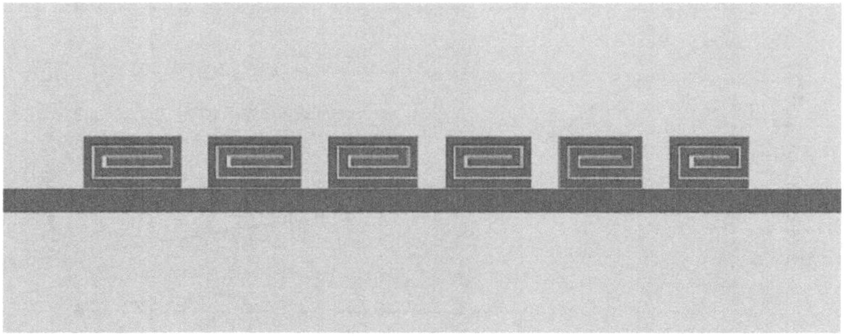

Fig. 3.26 Layout of 6-bit multiresonator in ADS Momentum 2008

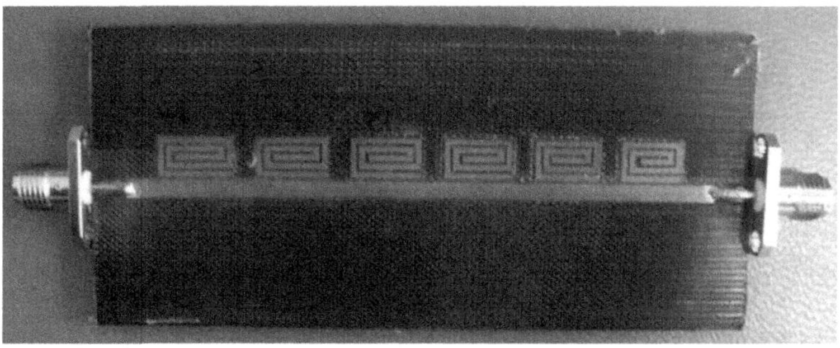

Fig. 3.27 Photograph of 6-bit multiresonator on Taconic TLX-0

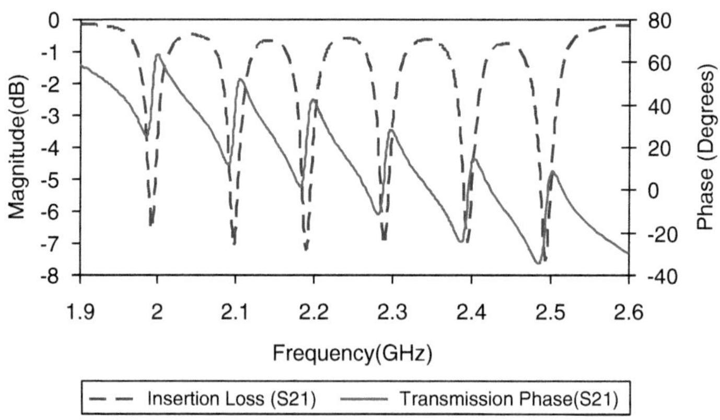

Fig. 3.28 Measured insertion loss and transmission phase of 6-bit multiresonator

Figure 3.28 shows the measured frequency response in both magnitude and phase of the 6-bit multiresonator. From Fig. 3.28, it is clear that at the resonant frequencies of individual spirals of the multiresonator there is a magnitude "dip" and phase "jump" in the magnitude and phase of the spectrum of the multiresonator. These properties are used to encode data into the spectrum using the multiresonator.

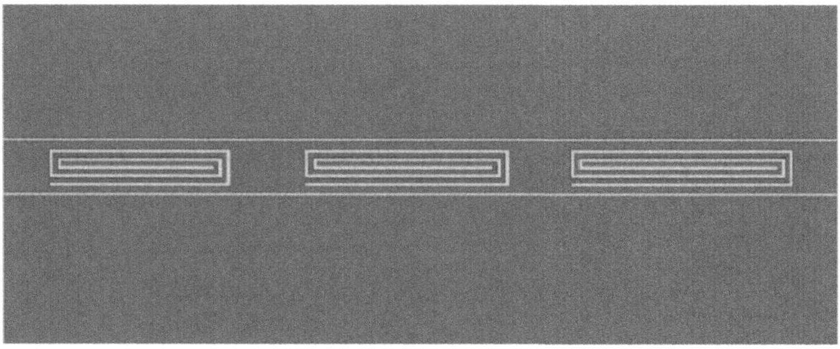

Fig. 3.29 Layout of coplanar waveguide (CPW) 3-bit multiresonator in ADS Momentum 2008

From the results shown in Fig. 3.28, we can see six distinct resonant nulls in the magnitude and six phase shifts due to the six spirals. Analogue 1:1 correspondence are binary data encoded by using the magnitude and phase of the spectral signature. The presence of a magnitude null (dip) and phase jump represents logic "0", while the absence of a magnitude null and phase jump at a particular frequency represents logic "1".

3.6.2 Multiresonator on Thin Flexible Laminate Using CPW Technology

In the previous section, the multiresonator was designed on thick PCB using microstrip technology. As shown in Sects. 3.4 and 3.5, CPW technology on thin flexible laminates is superior to microstrip technology. Hence, the design of a multiresonator on thin flexible laminate is performed on Taconic TF-290, which is 90 μm thin. Figure 3.29 shows the top view of the 3-bit multiresonator layout generated in ADS Momentum 2008. Figure 3.30 shows the photograph of the fabricated 3-bit multiresonator on TF-290 substrate. The 3-bit multiresonator consists of three spiral resonators cascaded within a 50 Ω CPW strip line. The spirals are etched in the CPW strip line. The CPW multiresonator provides three distinguishable resonances between 2 and 2.5 GHz. Each resonance is separated by approximately 200 MHz from each other. The resonant frequencies of the CPW spirals were placed further away in spectrum due to the larger bandwidth of the CPW spirals than microstrip spirals. In order to design the spirals at different frequencies, the length of each spiral has been varied so that the spiral's resonant frequency is tuned. The spiral resonators were placed 3 mm from each other.

Figure 3.31 shows the simulated frequency response in both magnitude and phase of the 3-bit multiresonator. From Fig. 3.31, it is clear that at the resonant frequencies of each spiral of the multiresonator there is a magnitude null and phase jump in the magnitude and phase of the spectrum of the CPW multiresonator. These distinct nulls and jumps in magnitude and phase, respectively, are interpreted as logic "0", while their absences at the resonant frequencies are interpreted as logic "1".

In the following section, the novel data-encoding technique for introducing and removing the resonances is presented.

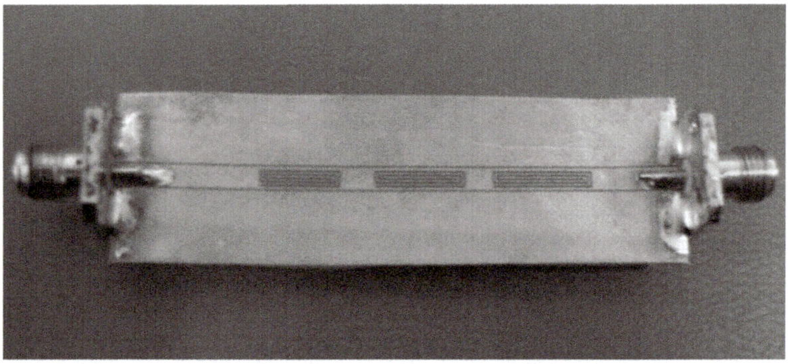

Fig. 3.30 Photograph of manufactured coplanar waveguide (CPW) 3-bit multiresonator on 90 μm Taconic TF-290 substrate

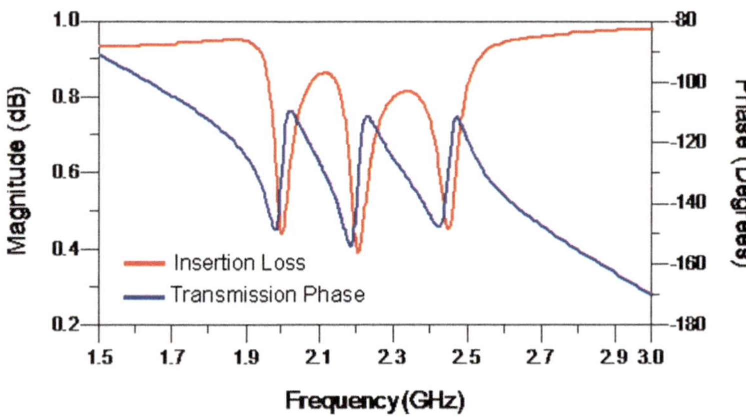

Fig. 3.31 Measured insertion loss and transmission phase of coplanar waveguide (CPW) 3-bit multiresonator

3.7 Encoding Data Using Novel "Spiral Shorting" Technique

It is necessary to encode data into the tag in order for the tag to have a unique ID. This can be done by introducing or removing the resonances of the multiresonator. Figure 3.28 shows the S-parameter measurements of a multiresonator, which gives a tag ID of 000000. In order to create a different ID, for example 101010, the resonances at 2.1, 2.3 and 2.5 GHz need to be removed. By removing the spiral, the resonance is removed as well. The other option is to short the turns of the spiral as shown in Figs. 3.32 and 3.33 (microstrip and CPW, respectively), thus shifting the resonance frequency of the spiral up where it will be of no significance. The shift of the resonant frequency with the shorting of the turns is shown in Fig. 3.34. The advantage of shorting turns in regards to removing the entire spiral from the layout

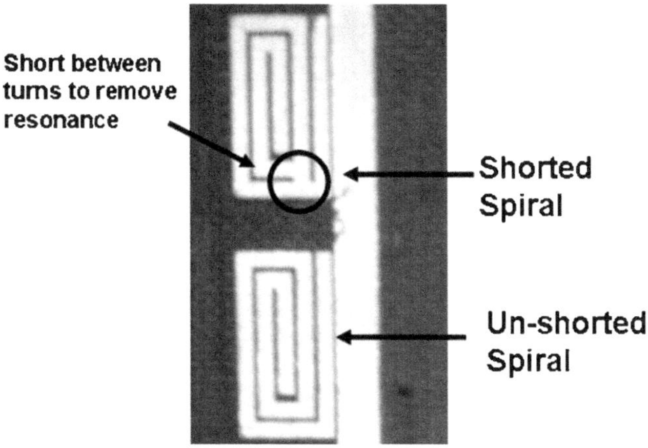

Fig. 3.32 Photograph of removing spiral resonances via spiral shorting for microstrip multiresonator

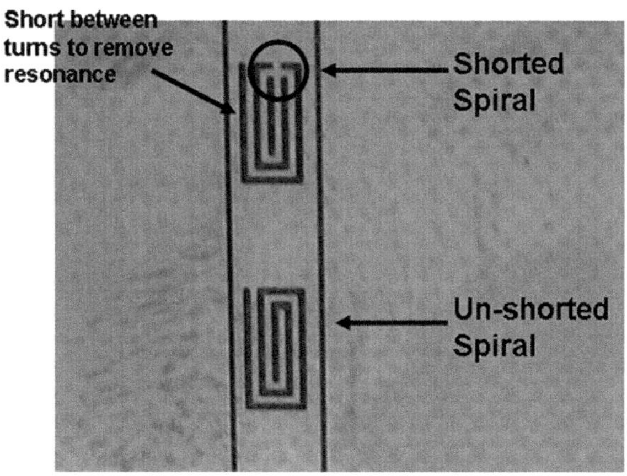

Fig. 3.33 Photograph of removing spiral resonances via spiral shorting for coplanar waveguide (CPW) multiresonator

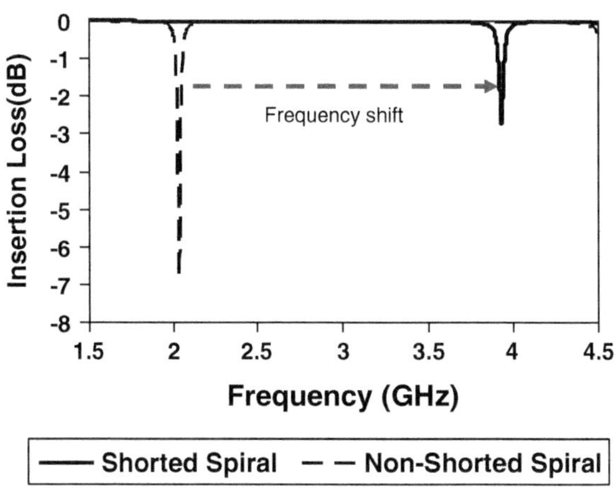

Fig. 3.34 Frequency shift of resonant frequency with short-circuited spiral

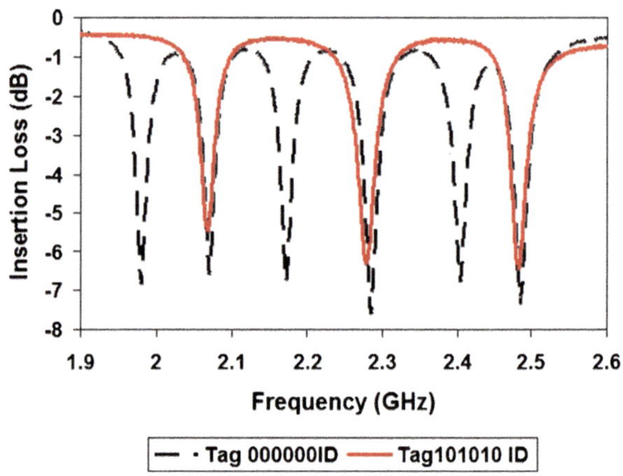

Fig. 3.35 Measured insertion losses of chipless tags with different spectral signatures

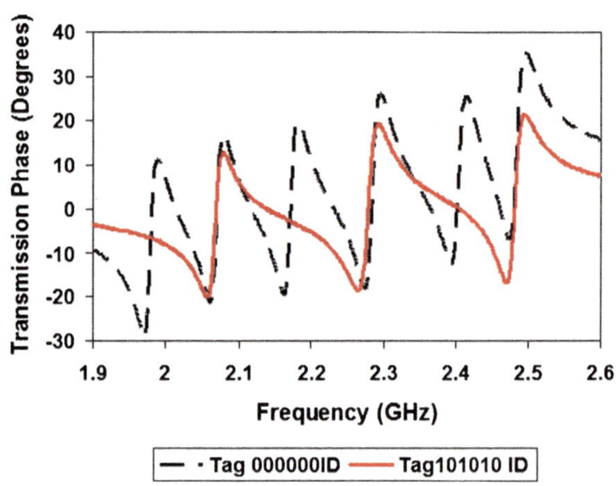

Fig. 3.36 Measured transmission phases of chipless tags with different spectral signatures

is the fact that it enables future printing techniques to preserve the layout with all of the spirals shorted and when encoding data the shorting can be removed via a laser or other etching technique. The frequency signatures of tags with different ID's are shown in Figs. 3.35 and 3.36.

3.8 Interference and Frequency Shifts of Cascaded Spiral Resonators

In the previous sections, the design of the multiresonator and data encoding using the spiral shorting method are presented. However, it is important to understand the influence and interference created amongst the cascaded spiral resonators. Figures 3.35 and 3.36 show the presence of a small frequency shift of the spiral resonance frequency.

Table 3.2 Resonant frequencies of microstrip spiral resonators under different design conditions

Spiral no.	All spirals active in multiresonator (GHz)	One spiral active in multiresonator	Single spiral (GHz)
1	1.993	1.994	1.994
2	2.098	2.099	2.099
3	2.191	2.190	2.191
4	2.291	2.291	2.290
5	2.396	2.395	2.395
6	2.494	2.493	2.494

Table 3.3 Resonant frequencies of CPW spiral resonators under different design conditions

Spiral no.	All spirals active in multiresonator (GHz)	One spiral active in multiresonator (GHz)	Single spiral (GHz)
1	1.997	1.998	1.998
2	2.204	2.201	2.2
3	2.449	2.445	2.445

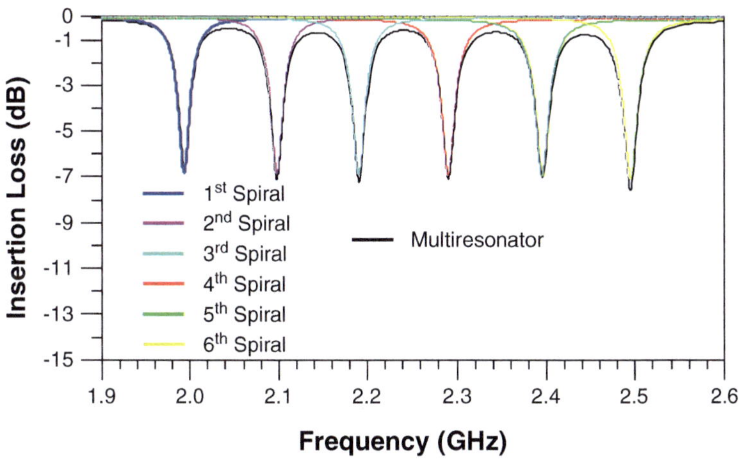

Fig. 3.37 Insertion losses of six spiral resonators in the multiresonator and alone

Tables 3.2 and 3.3 show the resonant frequencies of the microstrip and CPW multiresonator. Table 3.2 shows the resonant frequencies of the six microstrip spiral resonators, when they are cascaded in the multiresonator and all spirals are active, when they are cascaded in the multiresonator and only one spiral is active and others are shorted (inactive) and when they are alone. The same data is shown in Table 3.3 for three spirals in the CPW multiresonator. From Tables 3.2 and 3.3, it is clear that the frequency shifts are minimal (within a few MHz) and a completely acceptable. The influence of the spirals is shown in Figs. 3.37 and 3.38. From Figs. 3.37, it is clear that the frequency shifts of the spiral's resonant frequencies are minimal and that the spirals have minimal influence on each other.

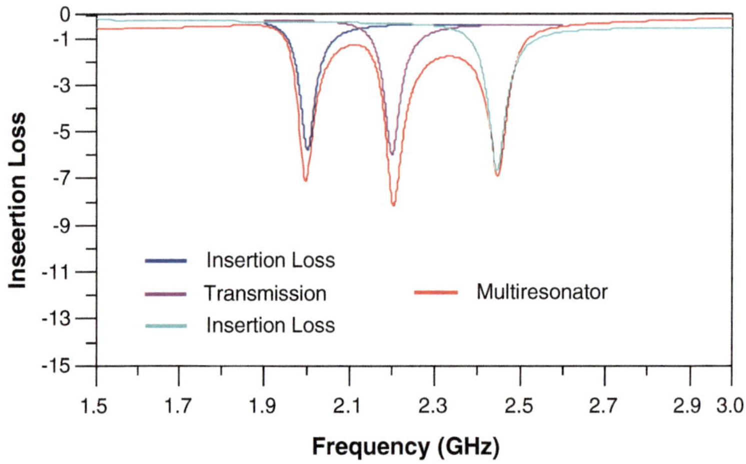

Fig. 3.38 Insertion losses of six spiral resonators in the multiresonator and alone

3.9 Conclusions

In this chapter, microstrip and CPW spiral resonators are investigated as high Q stop-band filters for the chipless tag multiresonators. In the first part of this chapter, the equivalent circuit and theoretical model of the microstrip spiral resonator is presented. The spiral resonator is modelled firstly with distributed LC components and then using coupled microstrip lines and micorstrip discontinuities.

The parametric study of the spiral resonator was performed by varying different design/layout parameters of the spiral resonator. Corresponding effects on the frequency responses of the spiral resonator due to the parametric study are also presented. A single spiral resonator was thus optimized and designed to produce a bandstop resonance at 2.37 GHz. The 3 dB bandwidth of the insertion loss is 20 MHz.

The microstrip spiral resonator was designed on thin flexible laminate TF-290. The performance of microstrip spiral resonators on flexible was investigated and yielded poor Q factor and unsatisfactory frequency response. Two different approaches of increasing the Q factor and optimizing the frequency response were performed: closed corner coupling and spiral repetition. Although these methods increased the insertion loss of the spiral at its resonant frequency, the bandwidth of the spiral resonators was dramatically increased (over 500 MHz).

The design of CPW spiral resonators etched into the strip line showed satisfactory performance on thin flexible laminate. The CPW spiral resonators exhibit comparable results with microstrip spiral resonators on thick PCB laminates. The advantages of CPW spirals in comparison with microstrip spiral resonators on thin flexible laminates are higher Q factor and single-sided layout. However, using CPW technology results in larger circuits due to the fact that cascading can be performed in 1D. Microstrip spiral resonators can be placed on both sides of the microstrip line, hence having smaller layout.

The multiresonator circuit is used for encoding data by the chipless tag. The multiresonator is designed as cascaded spiral resonators placed next to the microstrip line or etched in the CPW strip line using microstrip and CPW technology, respectively. The magnitude and phase of the frequency spectrum of the multiresonator circuit is investigated and shows that both can be used for data encoding. The spectral signature data encoding using phase is a fully novel concept and is intended to increase the reading range of the tag since phase is more resistant to noise than that for the magnitude.

A novel data-encoding technique call spiral shorting is presented. The encoding of data is performed by shorting the turns of the spiral in order to shift the spiral's resonant frequency outside the band of interest, hence resulting in no resonance at the particular frequency reserved for the spectral signature data bit. The presence of the resonance is treated as a logic "0", since it creates a magnitude null and phase jump, while its absence is treated as logic "1". The advantage of using this type of data encoding rather than removing the entire spiral is that there is little layout modification needed for data encoding and can be performed by a laser.

Different investigations were carried out to identify the mutual influence between cascaded spirals. The spiral resonators were placed in three different scenarios: (1) cascaded with all spirals active, (2) cascaded with only one spiral active while others are shorted (inactive) and (3) a single spiral resonator. The resonant frequencies of the spiral resonators in all three conditions resulted in minimal frequency shifts of only a few MHz, which are acceptable and reflect the robustness of the design.

The following chapter presents the design of UWB antennas for chipless RFID tag and RFID reader.

Chapter 4
Ultra Wideband Antennas

Abstract Chipless RFID tags require a significant amount of spectrum in order to encode their data. Hence, ultra wideband antennas seem like the obvious solution as the antenna of choice for the chipless RFID tag. The theory of operation and parametric study of circular disc monopole antennas printed on dielectric laminates is presented. Finally, an optimized monopole antenna for operating in the UWB spectrum is presented. The RFID reader antennas are also presented. The RFID reader requires a high gain antenna; hence, log periodic dipole antenna arrays on Taconic laminate are designed. The LPDA exhibits high gain, a directive radiation pattern and wide-operating bandwidth.

4.1 Introduction

In the preceding chapter, the theory and design of the spiral resonators for spectral signature encoding were discussed. In this chapter, the design of UWB antennas for the chipless tag and RFID reader are presented. The Federal Communications Commission (FCC) defined UWB antennas as any antenna that has an absolute bandwidth of no less than 500 MHz or fractional bandwidth of at least 0.2 [121].

The antennas used in this project are UWB due to the wideband spectrum required to encode spectral signature data. Based on the chipless RFID system specifications, it is clear that the antennas used for the tag and the reader are completely different. This is because the tag requires an omni-directinal antenna, while the reader circuit requires a high gain directional antenna.

Firstly, the theory of operation of UWB linearly polarized disc-loaded monopole antennas (here on known as only monopole antennas) is presented. Furthermore, the development of UWB monopole antennas for chipless tag on PCB and thin flexible laminate are presented in this chapter. Two types of technologies are used when designing the UWB monopoles on PCB and thin laminate: microstrip and CPW technologies, respectively.

S. Preradovic and N.C. Karmakar, *Multiresonator-Based Chipless RFID: Barcode of the Future*, DOI 10.1007/978-1-4614-2095-8_4,

The directional UWB antenna used for the RFID reader is a log periodic dipole antenna (LPDA) array. LPDAs are UWB high gain, linearly polarized directive antennas, which offer extended reading ranges for RFID readers.

4.2 Theory

In this section, the theory of operation of the disc-loaded monopole antennas and LPDAs is presented. The UWB monopoles are used as the Tx and Rx tag antennas, while the LPDAs are used as Tx and Rx reader antennas.

4.2.1 UWB Disc-Loaded Monopole Antennas for Chipless Tag

As in the case of any wireless communications system, the tag antenna plays a crucial role in the chipless RFID system. However, the choice of antenna and design is not a simple task and must meet certain criteria [122]. Firstly, the antenna should exhibit extremely large bandwidth in order to obtain many frequency signatures that are needed to encode large numbers of data bits. Secondly, the radiation pattern of the antenna should be as constant as possible over the entire frequency band of operation and should be omni-directional or very close to it. Thirdly, the antenna needs to be linearly polarized since orthogonal cross-polarization is to be used as an isolation mechanism between the interrogation signal sent from the reader and the tag's response signals received by the reader as mentioned in Chap. 2. Finally, the antenna has to exhibit a compact and fully printable (planar) layout in order to be printed on a paper-/plastic-based material.

Circular UWB monopole antennas have a simple layout and exhibit extremely large bandwidth and a figure-of-eight radiation pattern [123–127]. UWB dipole antennas have been reported by researchers with similar radiation properties and bandwidth as those found using UWB monopoles [128–130]. Both types of antennas are fully printable and at first seem suitable for deploying as the chipless RFID tag antenna. The disadvantage of UWB dipoles when compared to UWB monopoles is the fact that they require an impedance matching circuit and/or balun [131], which can increase the size and complexity of design of the tag. UWB monopole antennas need no impedance matching circuit or balun for their successful operation hence they are a preferred solution for the chipless RFID tag antenna.

The UWB characteristic of the monopole antenna is attributed to the overlapping of the antenna modes (resonances), which are closely distributed over the spectrum [132]. The operation principle of UWB disc monopole is shown in Fig. 4.1. The use of the higher order modes influences the monopole's radiation pattern in the *E* plane, which becomes distorted from a figure-of-eight radiation pattern in the fundamental mode. At the antenna's fundamental mode of operation, the wavelength of the transmitting/

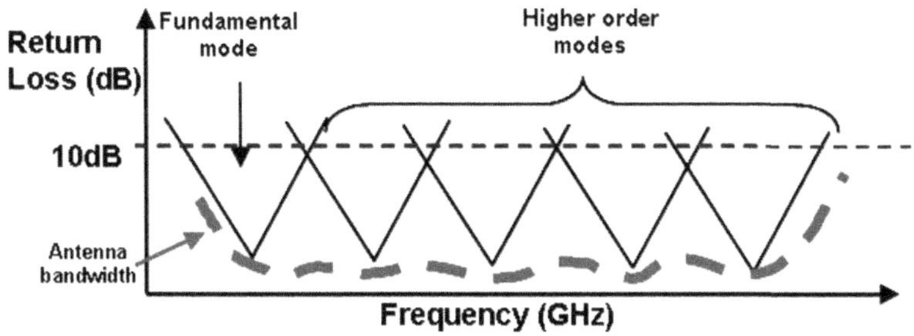

Fig. 4.1 UWB monopole operational principle

receiving signal is greater than the antennas dimensions and the antenna operates in an oscillating mode hence a standing wave is formed. As the frequency of the signal (and therefore operation) increases, the antenna starts operating in a hybrid mode of standing and travelling waves. At higher frequencies, the travelling waves are dominant since the wavelengths at these frequencies are quite smaller than the antenna structure. Therefore, due to the hybrid modes of antenna operation, it is possible to create an extremely wideband operating antenna at the expense of radiation pattern distortion at higher operating modes. These distortions are due to the surface current distribution in higher order modes [123, 124]. The *H*-plane radiation pattern remains constant throughout the operating band. This is due to the symmetry of the antenna's configuration along its axis of rotation.

4.2.2 *Log Periodic Dipole Antennas for Chipless Tag RFID Reader*

High gain RFID reader antennas with directive radiation patterns can significantly increase the RFID reader's reading range, and the number of tags that are interrogated as the antenna can provide spatial diversity with its narrow beamwidth. The LPDA consists of a sequence of side-by-side parallel linear dipoles, which form an antenna array. The LPDA has similarities in terms of layout and directivity with the Yagi-Uda array, but offers much larger bandwidths. The Yagi-Uda antenna elements do not follow any set patterns (lengths, element spacing separation, etc.), while the LPDA antenna elements increase logarithmically.

LPDAs have been designed as reader antennas in this book due to the fact that the proposed chipless RFID system needs wideband RFID reader antennas with directive pattern and high gain. The designed antenna presented in this section is designed for Gen1 and Gen2 RFID proof-of-concept readers, which operate between 2 and 2.5 GHz.

4.3 Design

In this section, the design steps of two types of UWB monopole antennas based on their feeding mechanisms: microstrip fed and CPW fed. Also, the empirical design steps of LPDA for RFID reader are presented.

4.3.1 Design of Microstrip Fed UWB Monopole

As discussed in the preceding section, the UWB monopoles need to be fully planar and printable. In this section, the design of UWB monopole antenna on thick PCB is performed. The substrate used is Taconic TLX-0 ($\varepsilon r = 2.45$, $h = 0.787$ mm, $\tan\delta = 0.0019$ and the antenna is fed using microstrip line [133–136].

The UWB monopole antenna layout with microstrip feed and parameters is presented in Fig. 4.2. As can be seen, the disc-loaded monopole antenna of disc radius *R* and the microstrip feed line of width Lfeed is etched on the top side of the laminate. The ground plane of width Wgnd and length Lgnd is etched on the bottom side of the laminate. The feedline is a 50 Ω microstrip line of width Wfeed. The antenna is fed with an SMA connector soldered at the end of the feed line.

The performance of the antenna relies entirely on its layout parameters. The two dominant design parameters are the radius of the patch *R* and the width of the ground plane Wgnd. The radius *R* determines the fundamental operating mode (first resonance) of the antenna, and therefore the low frequency cut off. The width of the ground plane Wgnd has significant contribution to the matching of the antenna and determines the high frequency cut off, hence the overall antenna bandwidth. The photograph of the UWB monopole antenna operating between 2 and 10 GHz is shown in Fig. 4.3.

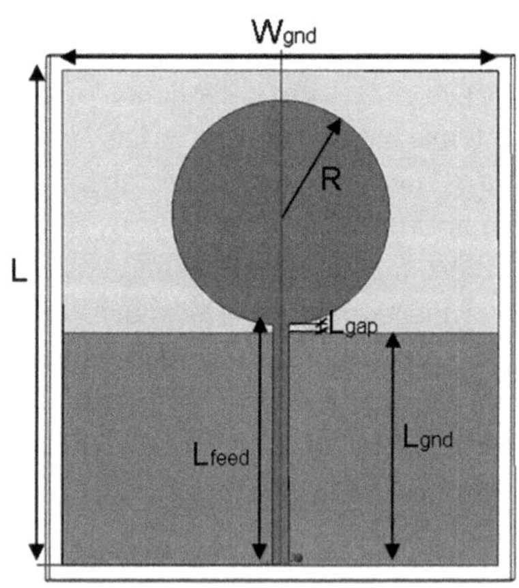

Fig. 4.2 UWB monopole antenna with microstrip feed and defined layout parameters

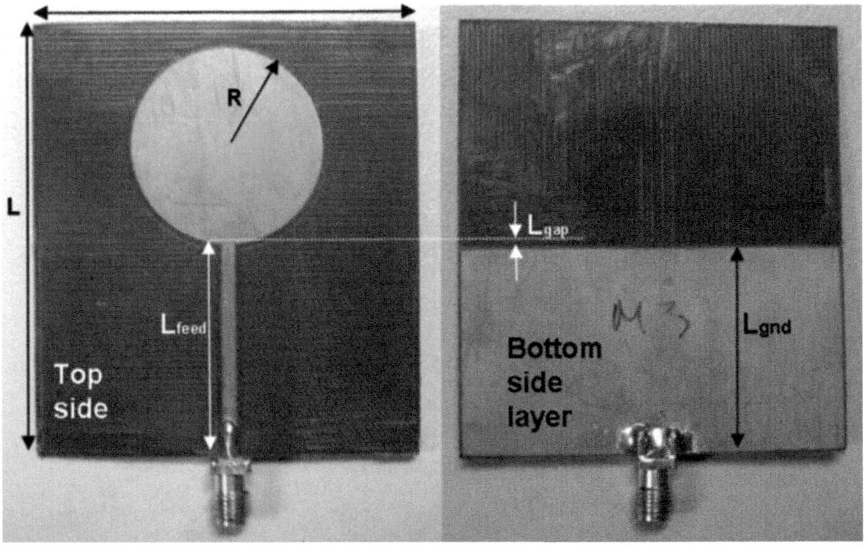

Fig. 4.3 Photograph of the manufactured UWB monopole with defined layout parameters (L = 66 mm, Wgnd = 60 mm, Lgap = 1 mm, Lgnd = 31 mm, Lfeed = 32 mm and R = 15 mm)

The UWB monopole antenna shown above was designed for a proof-of-concept tag operating from 2 to 2.5 GHz. However, the antenna operates in the UWB frequency spectrum, but it is too big in dimension to be used for a chipless tag on a banknote. The overall length of the antenna is 66 mm and width is 60 mm. Hence, the UWB antenna needs to be redesigned for working at a higher frequency range.

The new UWB tag antenna is designed in order to operate within the UWB (3.1–7 GHz) spectrum and with an almost omni-directional radiation pattern. The antenna was designed on Taconic TLX-0 using ADS 2008 Momentum and is shown in Fig. 4.4. The dimensions of the antenna are much smaller resulting in an antenna 36 mm in length and 20 mm wide. This type of antenna is suitable for the chipless tag with its dimensions.

4.3.2 Design of CPW Fed UWB Monopole

Following the design of the CPW multiresonator circuit on flexible laminate, the next step was to design a CPW fed UWB antenna. As presented in the previous sections, UWB monopole antennas are a suitable chipless RFID tag antenna with wideband return loss (RL) performance and almost omni-directional radiation patterns. Various UWB monopole antennas which incorporate CPW feeds have been reported in refs. [137–141]. However, the design of CPW fed UWB monopole antennas on flexible laminates has not been reported so far. Hence, a thorough parametric study of the antenna is performed.

Fig. 4.4 UWB monopole antenna with layout parameters (L=36 mm, W=20 mm, Lgnd=20 mm, Lgap=1 mm, Lfeed=21 mm and R=7 mm, substrate Taconic TLX-0 (εr=2.45, h=0.787 mm, tanδ=0.0019)

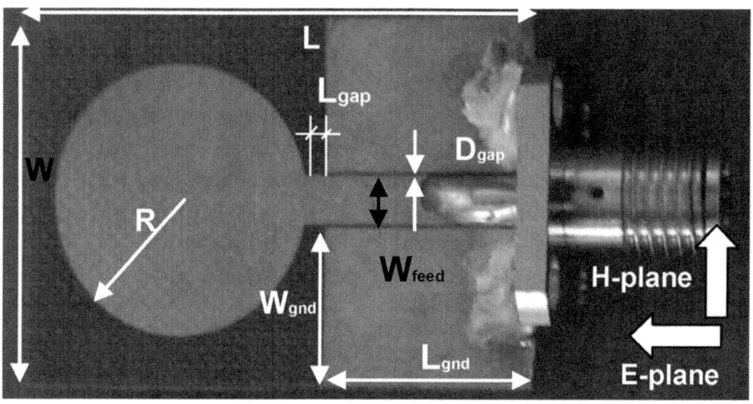

Fig. 4.5 Photograph of CPW fed UWB monopole (L=27 mm, W=18.8 mm, Lgnd=12 mm, Wgnd=8 mm, Dgap=0.15 mm, Lgap=2 mm, Wfeed=2.5 mm and R=7 mm, substrate Taconic TF-290 εr=2.9, h=0.09 mm, tanδ=0.0028)

The CPW disc monopole antenna is a single layer-metallic structure comprised of a copper disc with radius R and a 50 Ω CPW printed on the same side of the dielectric substrate Taconic TF-290 (εr=2.9, h=90 μm, tanδ=0.0028). The manufactured CPW monopole antenna with design parameters is shown in Fig. 4.5.

The parametric study of the UWB monopole antenna is shown in Figs. 4.6–4.9. The antenna's parametric study and design were performed using Agilent's ADS Momentum 2008 on 90 μm Taconic TF-290.

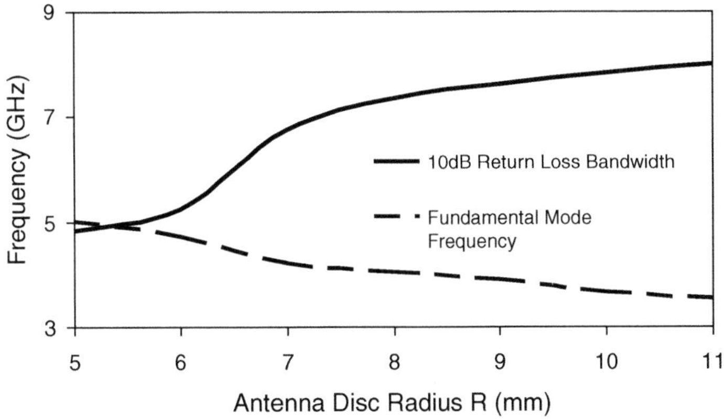

Fig. 4.6 Variation of antenna fundamental mode and 10 dB return loss bandwidth with *R* (Lgnd = 12 mm, Wgnd = 8 mm, Lgap = 2 mm; other parameters are as in Fig. 4.5)

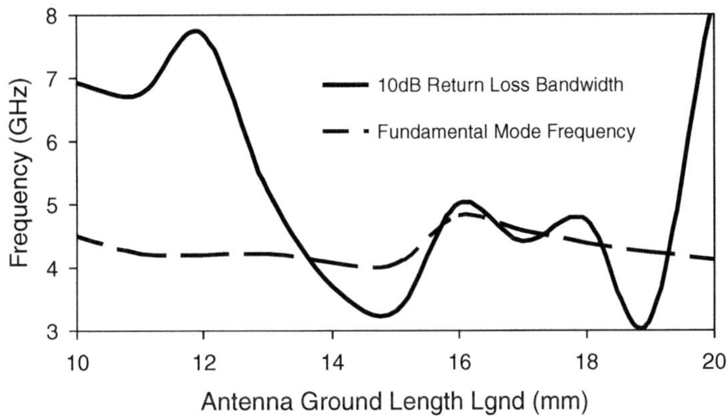

Fig. 4.7 Variation of antenna fundamental mode and 10 dB return loss bandwidth with Lgnd (*R* = 7mm, Wgnd = 8 mm, Lgap = 2 mm; other parameters are as in Fig. 4.5)

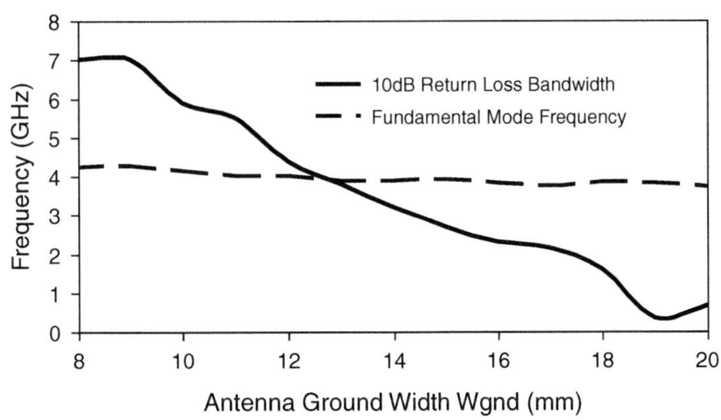

Fig. 4.8 Variation of fundamental mode and 10 dB return loss bandwidth with Wgnd (*R* = 7 mm, Lgnd = 12 mm, Lgap = 2 mm; other parameters are as in Fig. 4.5)

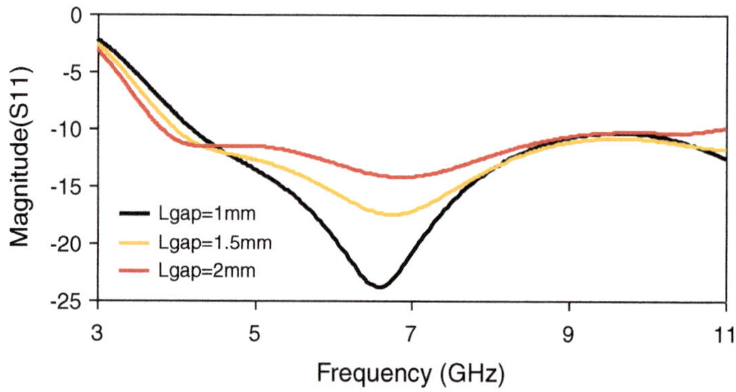

Fig. 4.9 Variation of antenna return loss and 10 dB bandwidth with Lgap ($R=7$ mm, Lgnd=12 mm, Wgnd=8 mm; other parameters are as in Fig. 4.5)

Figure 4.6 indicates that for $R=7$ mm the fundamental radiating mode of the antenna is around 4.2 GHz with a 10 dB bandwidth greater than 6.5 GHz, which covers the tag's operating frequency range (5–10.7 GHz) with small physical dimensions. Figure 4.7 shows that the 10 dB antenna bandwidth remains above or around 6.5 GHz for shorter (<12 mm) and much longer (>20 mm) ground plane lengths. When designing the tag, the ground plane length is longer than 20 mm, hence the RL bandwidth is greater than 7 GHz, which satisfies the system's operating requirements. From Fig. 4.8, it is clear that the variation of the ground width does not influence the resonant frequency, but significantly influences the 10 dB RL bandwidth of the antenna. The optimized value for the antenna's ground width is 8 mm since it provides the largest antenna bandwidth. Figure 4.9 shows that the optimized value of Lgap=2 mm is the optimal solution for the best matching of the antenna over the defined bandwidth.

Figure 4.10 shows the simulated return losses of the Tag Rx and Tag Tx antennas integrated within the tag. The RL simulations were performed in order to test the performance of the monopoles once they are integrated with the multiresonating circuit hence resulting in a greater and non-symmetrical monopole ground plane. The RL is still satisfactory (below −10 dB) over the desired region from 5 to 11 GHz where the tag data is encoded. The radiation pattern distortions are minimal due to the main surface current distribution being around the gap between the antenna ground plane and circular disc.

4.3.3 Design of Log Periodic Dipole Antennas

An LPDA design needs three design parameters [142]: geometric ratio τ, angle factor α and spacing factor σ as shown in Fig. 4.11.

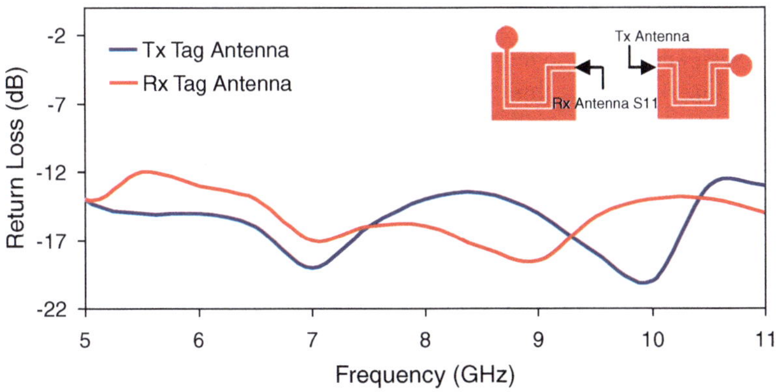

Fig. 4.10 Simulated Tx and Rx tag antenna return losses once integrated into the chipless tag circuit

Fig. 4.11 Layout of log periodic dipole array with layout-defined parameters

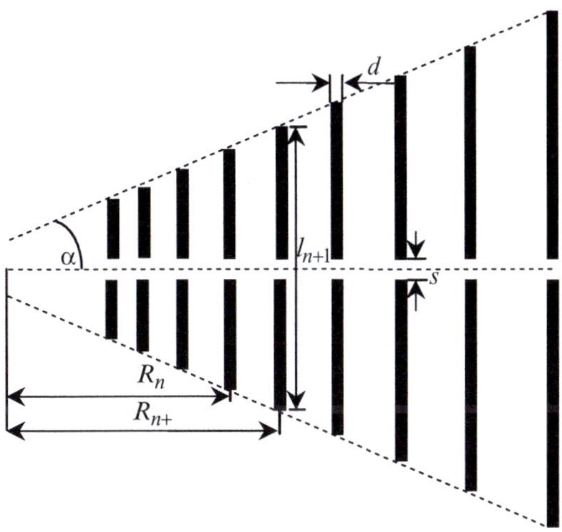

By defining the desired directivity, the geometric ratio τ and spacing factor σ may be easily found from [142]. The geometric ratio determines the length and resonant frequency of each dipole element using

$$\tau = \frac{l_{n+1}}{l_n} = \frac{f_{rn+1}}{f_{rn}}, \tag{4.1}$$

where f_{rn} and f_{rn+1} are the resonant frequencies of the n-th and $n+1$-st dipole element, respectively. The spacing factor σ defines the separation between the dipole elements in order to achieve desired bandwidth and gain and is defined as

$$\sigma = \frac{R_{n+1} - R_n}{2l_{n+1}}, \tag{4.2}$$

where R_n and R_{n+1} are the separation between the n-th and $n+1$-st dipole element, respectively, and l_{n+1} is the length of the two dipole arms of the $n+1$-st element as shown in Fig. 4.11. After this, the angle factor may be calculated using the following equation:

$$\alpha = \tan^{-1}\left[\frac{1-\tau}{4\sigma}\right]. \tag{4.3}$$

Based on these three design parameters, the LPDA is designed with the layout shown in Fig. 4.11.

The antenna bandwidth is defined as

$$B_s = \frac{f_{\text{stop}}}{f_{\text{start}}} \times (1.1 + 7.7(1-\tau)^2 \cot\alpha), \tag{4.4}$$

where f_{stop} is the high frequency limit and f_{start} is the low frequency limit. The overall length of the antenna is defined as

$$L = \frac{\lambda_{\max}}{4}(1 - \frac{1}{B_s})\cot\alpha, \tag{4.5}$$

where $\lambda_{\max}$ is the largest wavelength defined as

$$\lambda_{\max} = 2l_{\max} = \frac{v}{f_{\min}}. \tag{4.6}$$

The number of dipole elements that need to be used for the LPDA is defined as

$$N = 1 + \frac{\ln B_s}{\ln(1+\tau)}. \tag{4.7}$$

The characteristic impedance Z_a of the dipole elements is defined as

$$Z_a = 120\left[\ln\frac{l_n}{d_n} - 2.25\right], \tag{4.8}$$

where l_n and d_n are the length and width of the n-th dipole element, respectively.

The design curve shown in ref. [142] shows the relationship between Z_a / R_{in} and Z_0 / R_{in}, where Z_0 is the antenna input impedance and R_{in} is the desired input impedance (in our case 50 Ω). Based on this curve, we can set the necessary impedance of the dipole arms Z_a in order to set the ratio Z_0 / R_{in} to 1 in order to achieve a match to 50 Ω. Finally, the spacing between the feed lines is found from:

$$s = d \times \cosh\left(\frac{Z_0}{120}\right). \tag{4.9}$$

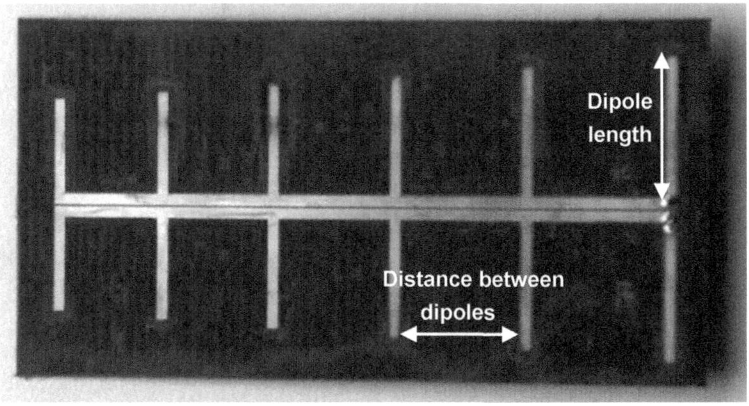

Fig. 4.12 Photograph of the reader LPDA with defined layout parameters (substrate Taconic TLX-0 $\varepsilon r = 2.45$, $h = 0.787$ mm, $\tan\delta = 0.0019$)

Table 4.1 Log periodic dipole array dimensions

Dipole length (cm)	Distance between dipoles (cm)
3.22	3.22
2.99	2.99
2.78	2.78
2.58	2.58
2.4	2.4
2.24	2.24

Further optimization of the LPDA is performed using CAD tools, such as ADS, CST, Ansoft HFSS, etc.

The overall design process of the reader LPDA may be summarized as:

1. Determine σ and τ from given directivity (dB).
2. Determine α using (4.3).
3. Determine antenna bandwidth B_s using (4.4).
4. Calculate L and N from (4.5) and (4.7), respectively.
5. Determine Z_a / R_{in} and Z_0 / R_{in} using [99].
6. Calculate spacing between the feed lines s from (4.9).
7. Perform LPDA optimization using CAD tools.

Optimization of the LPDA was done using CST Microwave Studio 2008. Figure 4.12 shows the photograph of the designed LPDA on Taconic TLX-0. The layout dimensions of the LPDA are given in Table 4.1.

4.4 Results

In this section, the measured results of the CPW fed and microstrip fed UWB monopole antennas and LPDA are presented. The results include RL, gain, co-polar and cross-polar radiation pattern measurements.

4.4.1 *Microstrip Fed UWB Monopole Antenna Results*

The antenna is measured using Agilent's PNA E8136A with one port calibration from 1 to 10 GHz. The measured RL vs. frequency of the monopole antenna is shown in Fig. 4.13. As can be seen in Fig. 4.13, the antenna exhibits 10 dB RL bandwidth starting from 2 GHz to beyond 10 GHz.

Figures 4.14 and 4.15 show the radiation patterns of the UWB tag antenna at 2 and 2.5 GHz measured inside an anechoic chamber, respectively. As can be seen, the minimum cross-polar level is −25 dB (160° at 2 GHz vertical plane) and the maximum is −7 dB (90° at 2.5 GHz vertical plane) smaller than the co-polar component. Therefore, the measured radiation patterns ensure polarization purity. The measured antenna gain was ~1 dBi across the desired frequency band. The suppression of the cross-polar component is essential and can contribute significantly to robust reading results.

As mentioned in the design Sect. 4.3.1, a new monopole antenna with smaller dimensions was designed. The tag antenna measurements were performed in the anechoic chamber at the Department of Electrical and Computer Systems Engineering (ECSE) at Monash University.

Figure 4.16 shows the RL vs. frequency of the monopole antenna. The fundamental mode (at 3.5 GHz) and higher order modes of the UWB monopole antenna are clearly visible in Fig. 4.16. As mentioned in the theory of the antenna, the UWB

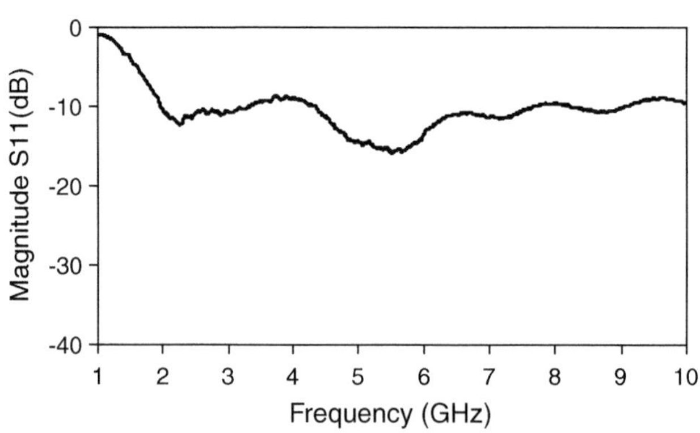

Fig. 4.13 Measured return loss of chipless RFID tag UWB monopole antenna

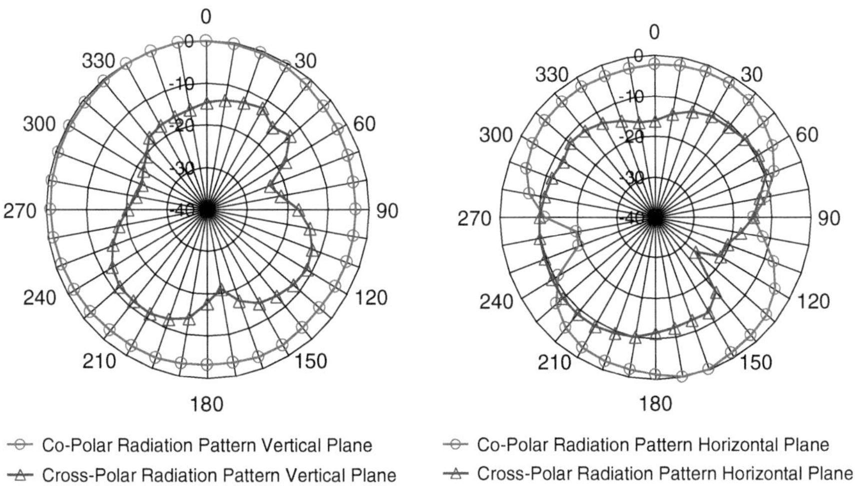

Fig. 4.14 Measured co-polar and cross-polar radiation patterns of UWB monopole at 2 GHz

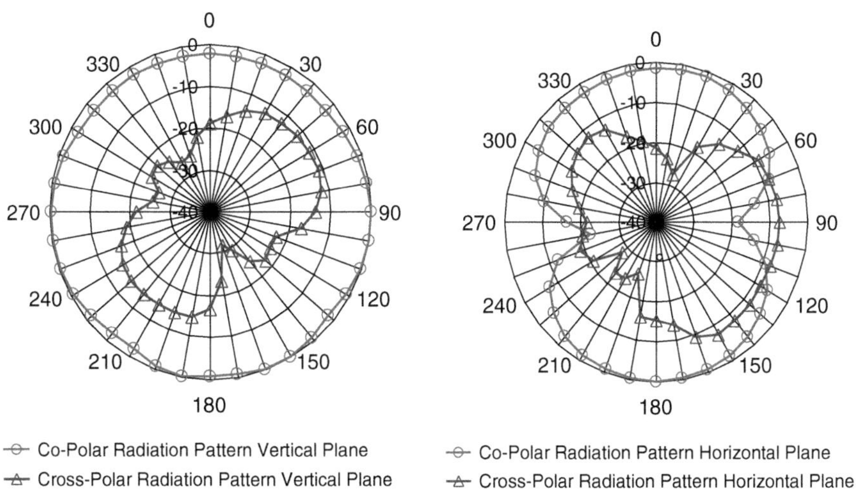

Fig. 4.15 Measured co-polar and cross-polar radiation patterns of UWB monopole at 2.5 GHz

monopole antennas rely on the overlapping of their modes in order to achieve extremely large bandwidth. Hence, radiation pattern changes are expected at frequency ranges which are not within the monopole's fundamental mode of operation. The antenna co-polar and cross-polar radiation patterns from 3 to 7 GHz (in steps of 1 GHz) in both horizontal (*E* plane) and vertical planes (*H* plane) are presented in Figs. 4.17–4.21, respectively. As can be seen in the figures, the *H*-plane patterns are almost isotropic due to the symmetry of the antenna structure in the *H* plane.

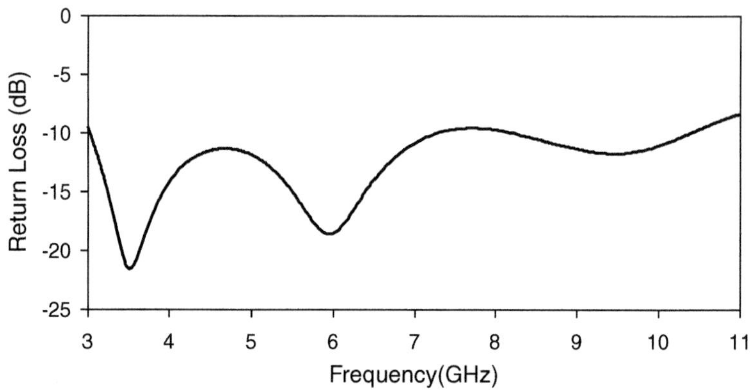

Fig. 4.16 Measured UWB chipless tag antenna return loss

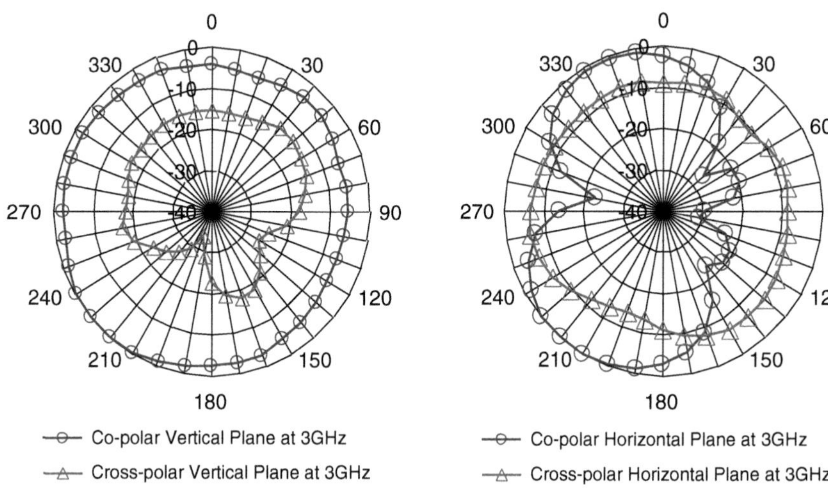

Fig. 4.17 Measured tag antenna radiation patterns at 3 GHz

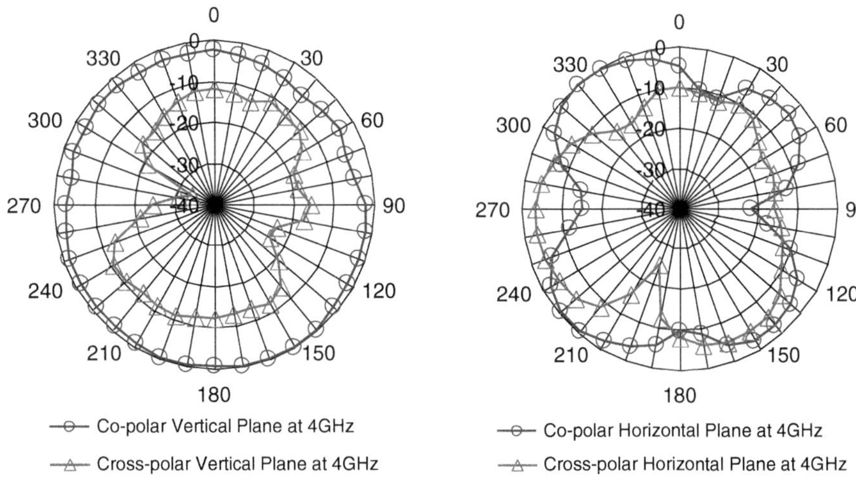

Fig. 4.18 Measured tag antenna radiation patterns at 4 GHz

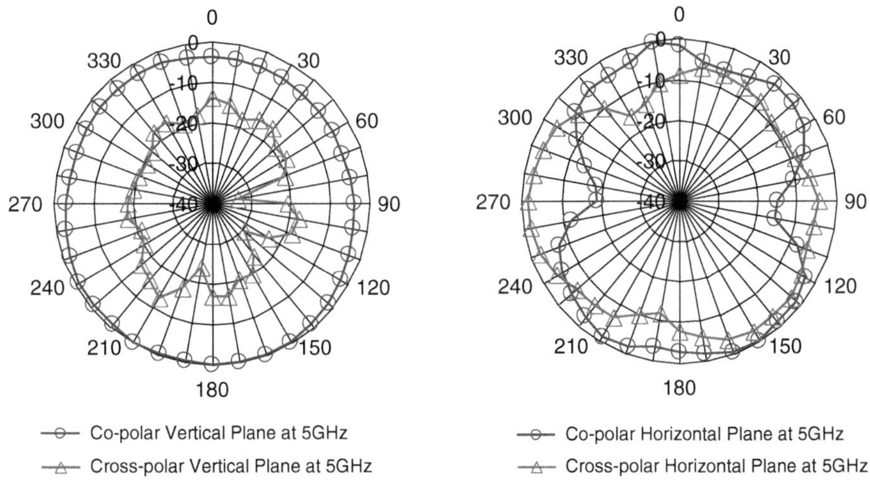

Fig. 4.19 Measured tag antenna radiation patterns at 5 GHz

The measured peak gain of the UWB monopole tag antenna was measured in the anechoic chamber and is shown in Fig. 4.22. The gain of the antenna increases with frequency which is as expected.

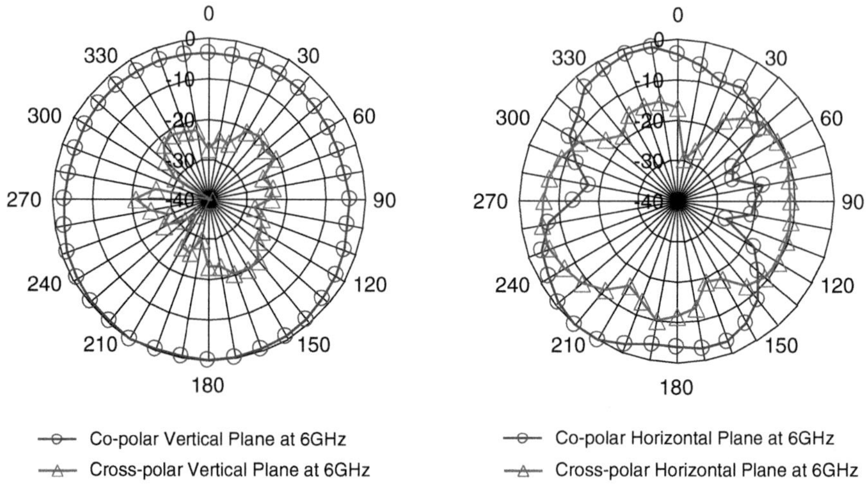

Fig. 4.20 Measured tag antenna radiation patterns at 6 GHz

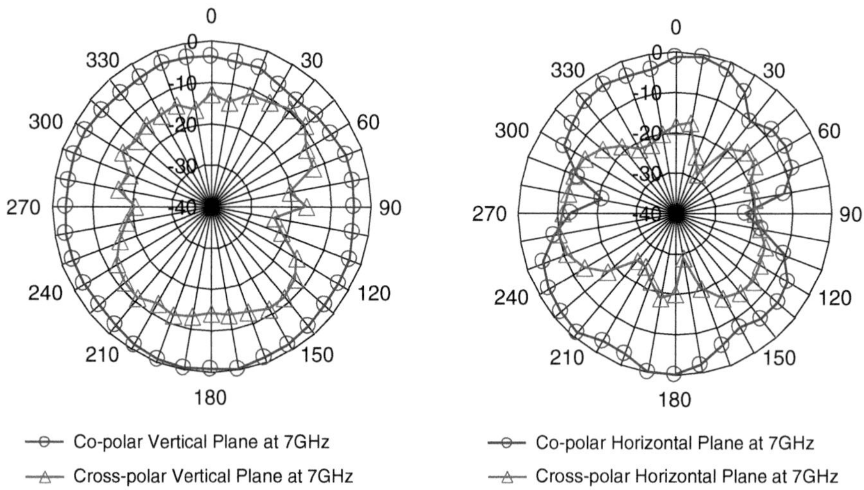

Fig. 4.21 Measured tag antenna radiation patterns at 6 GHz

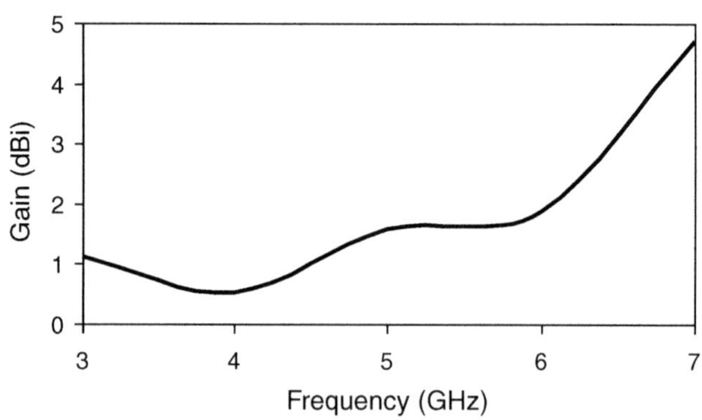

Fig. 4.22 Measured tag antenna radiation patterns at 7 GHz

4.4.2 CPW Fed UWB Monopole Results

The measured antenna RL vs. frequency is shown in Fig. 4.23. The antenna yields UWB operation with greater than 10 dB RL from 5 to 11 GHz.

Figure 4.23 shows multiple operating modes of the UWB monopole antenna. In order to achieve extremely large bandwidth, UWB monopole antennas rely on the overlapping of their modes. Hence, the changes in radiation patterns are expected with frequencies outside the monopole's fundamental mode of operation of around 4.2 GHz. The antenna co-polar and cross-polar radiation patterns from 5 to 10 GHz (in steps of 1 GHz in an anechoic chamber) in both *E* and *H* planes (as shown in Fig. 4.5) are presented in Figs. 4.24–4.29, respectively.

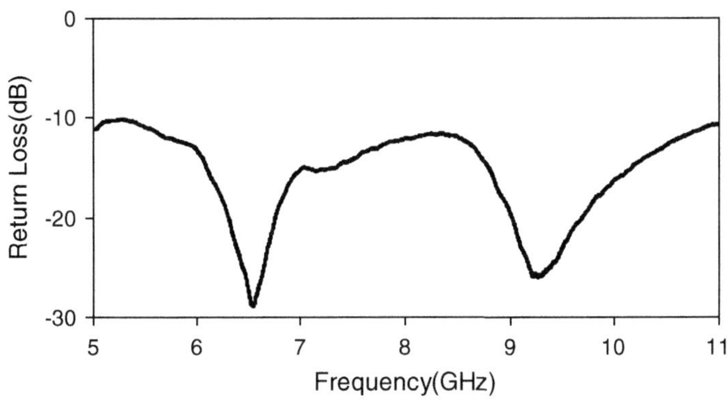

Fig. 4.23 Measured return loss of chipless RFID tag UWB monopole antenna

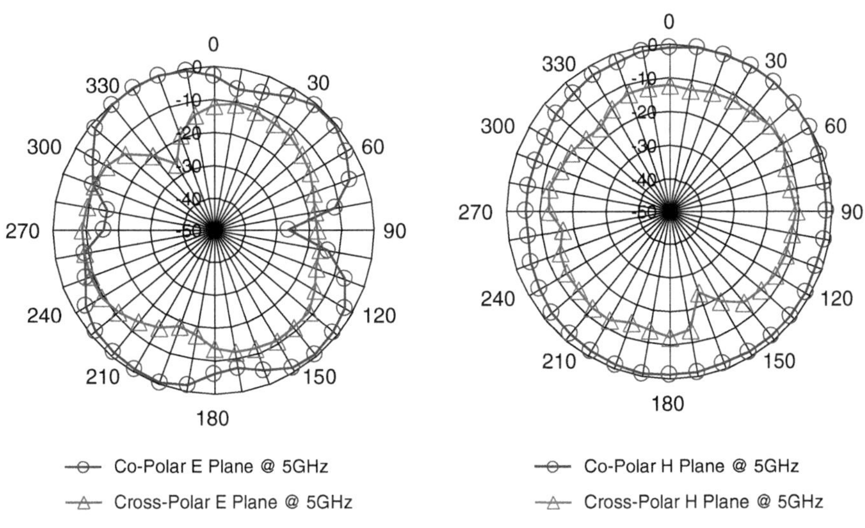

Fig. 4.24 Measured co-polar and cross-polar radiation patterns of UWB monopole at 5 GHz

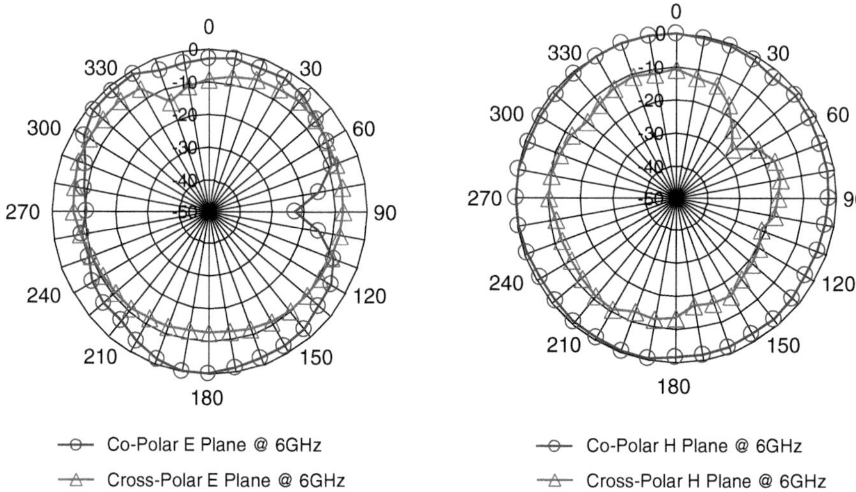

Fig. 4.25 Measured co-polar and cross-polar radiation patterns of UWB monopole at 6 GHz

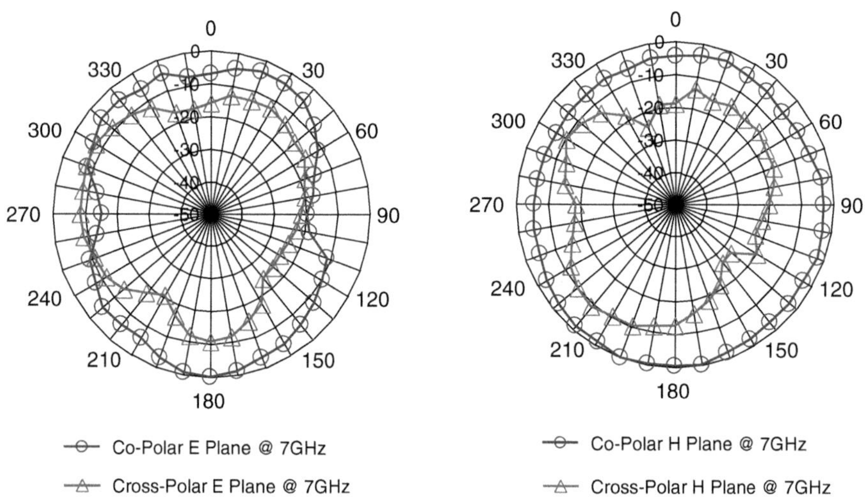

Fig. 4.26 Measured co-polar and cross-polar radiation patterns of UWB monopole at 7 GHz

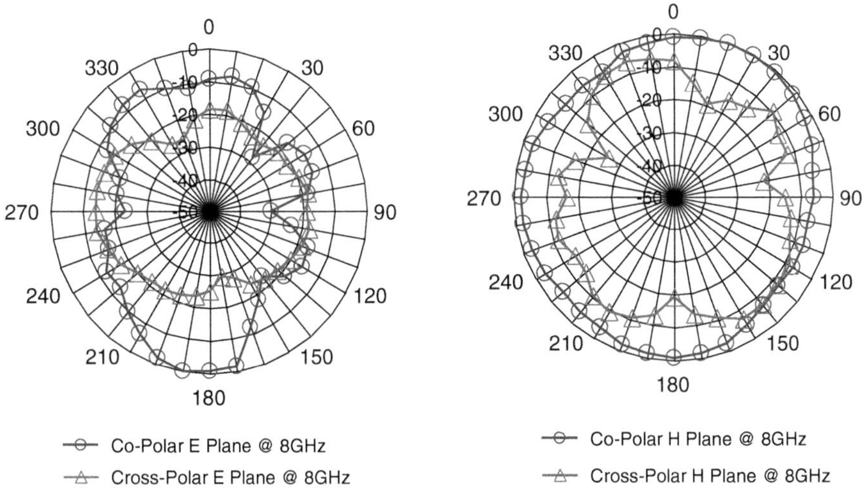

Fig. 4.27 Measured co-polar and cross-polar radiation patterns of UWB monopole at 8 GHz

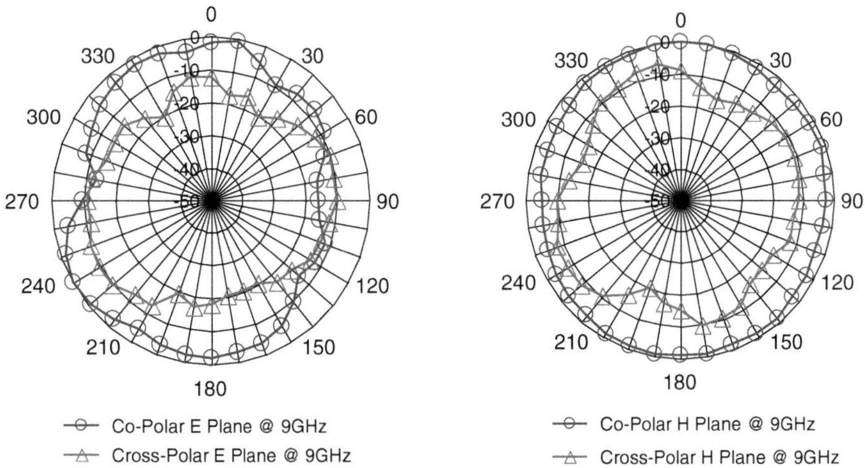

Fig. 4.28 Measured co-polar and cross-polar radiation patterns of UWB monopole at 9 GHz

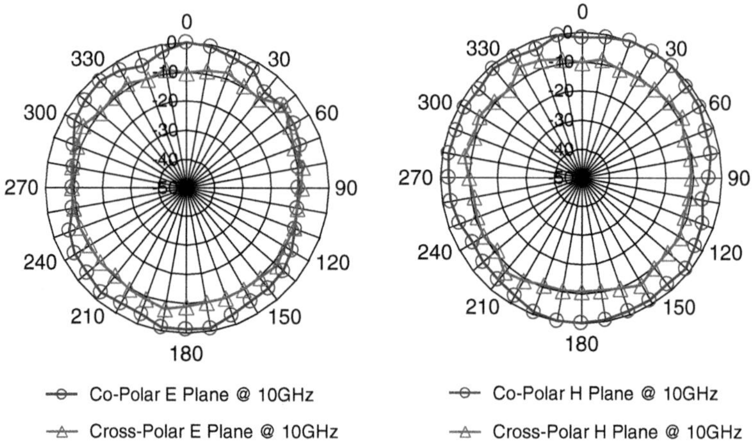

Fig. 4.29 Measured co-polar and cross-polar radiation patterns of UWB monopole at 10 GHz

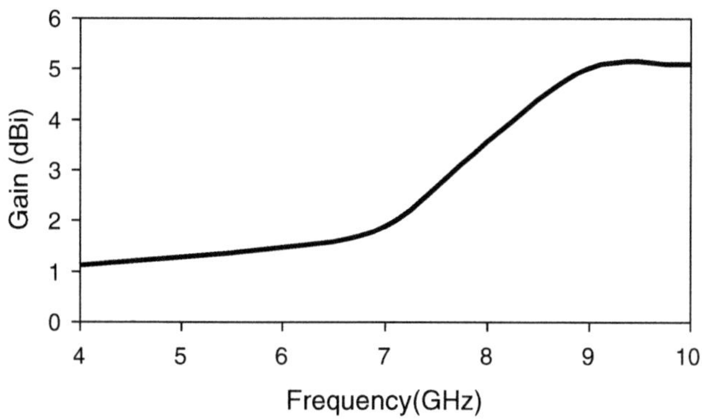

Fig. 4.30 Measured peak gain of chipless RFID tag UWB monopole antenna

The tag antennas show good cross-polar component suppression (at least in the order of 10 dB average), which is essential for robust readings and isolation between the interrogation signal and encoded signal. Figure 4.30 shows the measured peak gain vs. frequency of the tag antenna. The gain of the antenna increases with frequency.

The following section presents the design of directional UWB reader antennas for RFID reader applications.

4.4.3 Log Periodic Dipole Antenna Results

The measured RL and gain vs. frequency, co-polar and cross-polar radiation patterns in both vertical and horizontal planes in near- and far-field conditions are shown in Figs. 4.31 and 4.32, respectively. The antenna measurements were performed in an anechoic chamber. The near-field radiation patterns are of particular interest due to the fact that the proposed tag is read mostly in the near-field of the

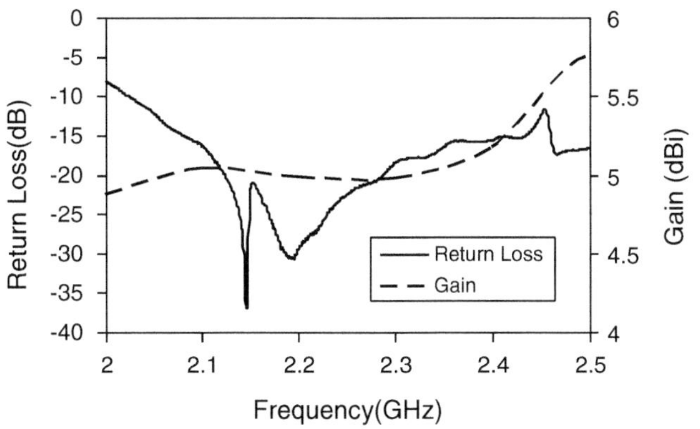

Fig. 4.31 Measured return loss and gain of LPDA reader antenna

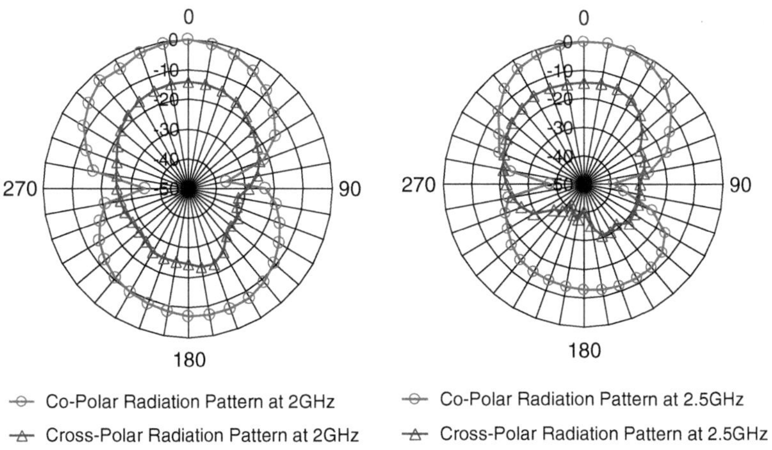

Fig. 4.32 Measured near-field co-polar and cross-polar radiation patterns of LPDA reader antenna in horizontal plane (18 cm distance)

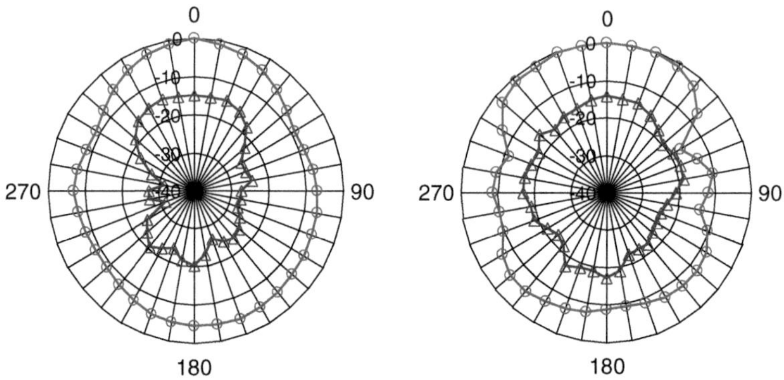

Fig. 4.33 Measured near-field co-polar and cross-polar radiation patterns of LPDA reader antenna in vertical plane (18 cm distance)

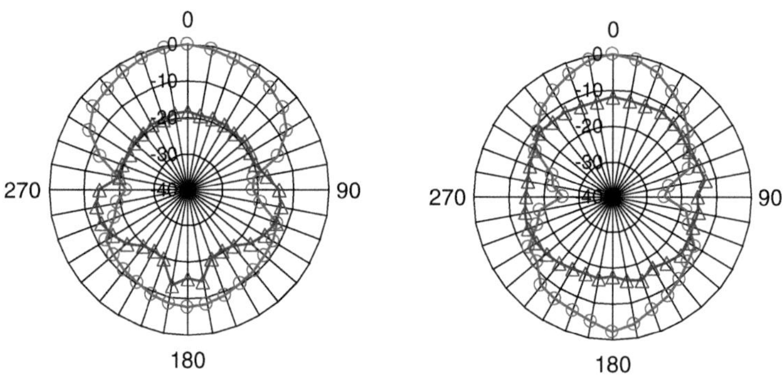

Fig. 4.34 Measured far-field co-polar and cross-polar radiation patterns of LPDA reader antenna in horizontal plane (86 cm distance)

LPDA reader antenna. As can be seen from Figs. 4.32 and 4.33, the polarization purity in the near-field region is still intact with more than 10 dB cross-polar levels.

The LPDA is a preferred candidate for the reader antenna to the UWB monopoles due to more directive radiation pattern, which creates a stronger line-of-sight (LOS) component, higher gain and low cross-polar components (Figs. 4.34 and 4.35).

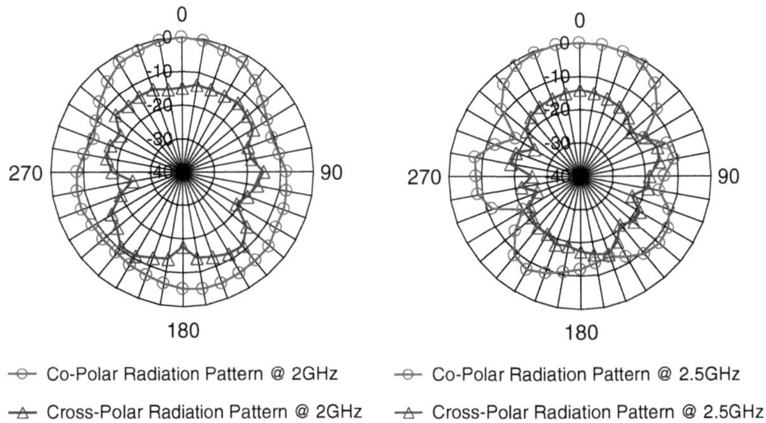

Fig. 4.35 Measured far-field co-polar and cross-polar radiation patterns of LPDA reader antenna in vertical plane (86 cm distance)

4.5 Conclusions

Three different types of antennas have been investigated in this chapter. They include UWB disc-loaded monopole antennas in both microstrip and CPW for chipless tag and LPDA for the RFID reader. The goal of these investigations has been to achieve a tag antenna with wideband operation and omni-directional radiation pattern and a directive wideband high gain antenna for the RFID reader, respectively.

The monopole antenna was designed using two different feeding methods. The first method used was the microstrip feed line. A proof-of-concept monopole antenna operating from 2 to 2.5 GHz was designed and then a UWB monopole operating from 3 to 7 GHz was designed. Both antennas were tested in an anechoic chamber for RL, antenna radiation patterns (co-polar and cross-polar in both E and H planes) and gain. The measured return losses of the antennas were above 10 dB in the desired operating ranges, which confirmed the successful matching of the antennas to a $50\,\Omega$ impedance at the desired frequency band. The antenna radiation pattern measurements showed that the antennas have an almost omni-directional co-polar component and a reasonable cross-polar component suppression of 7 dB in both planes. The cross-polar component suppression is particularly important due to the fact that the interrogation signal and tag signal are cross-polarized which minimizes cross-talk.

The design of the second type of monopole antenna was on thin flexible TF-290 laminate. The monopole antenna was fed by a CPW $50\,\Omega$ strip line. This was necessary since the multiresonator of the tag on flexible laminate is designed using CPW technology. However, little information on CPW UWB monopole antenna design on flexible laminate was available in literature, hence a comprehensive parametric study of CPW monopole was conducted by varying the antennas layout properties. From the parametric study, it was clear that the resonant frequency of the antenna is mainly determined by the size of the antenna's circular disc (or patch) and that the width and length of the ground plane influence the matching of the antenna to 50 Ω. The antenna was designed to operate between 5 and 10.7 GHz with an almost omni-directional radiation pattern and smaller gain due to the radiation pattern requirements. The antenna co-polar and cross-polar components of the radiation patterns in both planes were measured in an anechoic chamber along with the gain. The antenna radiation patterns were almost omni-directional with some distortions in the *E* plane due to higher order modes operating at higher frequencies.

The LPDA was designed as a high gain UWB antenna for the proof-of-concept reader operating between 2 and 2.5 GHz. The design steps of the LPDA entailed the calculation of the dipole arms and separation distances between them. The final tuning and optimization of the LPDA was done using CAD tools. The LPDA was designed on Taconic TLX-0 substrate using CST Microwave Studio 2008. The measured RL of the antenna was measured to be above 10 dB from 2 to 2.7 GHz. The radiation pattern and gain measurements of the LPDA were performed in the anechoic chamber. The measured gain was between 4.7 dBi at 2 GHz and 6.5 dBi and 2.7 GHz. The LPDA radiation pattern was measured in both planes in near- and far-field conditions. The radiation pattern confirmed the directive radiating nature of the LPDA and that the cross-polar component suppression above 10 dB.

The following chapter will present the integration of the UWB monopole antenna and multiresonator into a chipless RFID tag and its field trials.

Chapter 5
Chipless RFID Tag

Abstract This chapter focuses on the integration of the antenna and multiresonator structures presented in the previous chapters to form the chipless RFID tag. The design of three chipless RFID tag prototypes is presented. The first tag that is designed is a proof-of-concept tag with 6 bits of data. The second tag is a 35-bit tag designed in microstrip technology. The third tag is single layered (CPW technology) with 23 bits of data developed on thin flexible laminate. All three tags are tested in an anechoic chamber to prove that they are functional and operational.

5.1 Introduction

The preceding chapters presented the design of spiral resonators and ultra-wide band (UWB) monopole antennas using microstrip and CPW technology. The spiral resonators are the data-encoding devices of the proposed chipless RFID tag. The transmitting (Tx) and receiving (Rx) antennas of the tag are used to receive and transmit the interrogation signal from the reader and encoded tag signal, respectively.

The integration of the UWB monopole antennas and multiresonating circuit which form the complete chipless RFID tag is presented in this chapter. Firstly, the operating principle of the chipless tag is presented. The tag is fully passive and hence requires an external source of electromagnetic signals in order to encode the data. The external source of EM signals comes from the reader, which transmits a continuous wave interrogation signal to the tag. The tag receives the signal and encodes its spectral signature using the multiresonator. The encoded tag signal is then sent back to the reader by the tag's transmitting antenna. The reader further process the returned data and displays the ID code of the tag.

The development of the chipless RFID tag is presented in three stages. Firstly, the 6-bit chipless tag operating from 2 to 2.5 GHz is presented. It is a first proof-of-concept tag on a standard microstrip laminate Taconic TLX-0. The successful design and testing of the 6-bit tag provided motivation for the design of a UWB 35-bit tag

S. Preradovic and N.C. Karmakar, *Multiresonator-Based Chipless RFID: Barcode of the Future*, DOI 10.1007/978-1-4614-2095-8_5,

that operates from 3.1 to 7 GHz. Both the 6-bit tag and the 35-bit tag are designed on 787-μm thick Taconic TLX-0. Finally, the design of a 23-bit chipless tag on 90-μm thin flexible laminate is presented. All three tags are tested inside and outside an anechoic chamber to confirm their successful operations.

5.2 Chipless RFID Tag Operating Principle

In this section, the operating principle of the chipless tag is presented. The fundamentals of the working principle of the chipless tag and reader system were presented in Chap. 1 (Proposed Chipless RFID System). The signal flow diagram of the proposed chipless tag is shown in Fig. 5.1.

The chipless tag encodes data in the frequency spectrum, thus encoding the spectrum with its unique spectral signature. The spectral signature is obtained by the RFID reader by means of interrogating the tag by a multifrequency signal. The tag encodes its spectral signature into the interrogation signal spectrum using a multiresonating circuit, which is a multistop band filter. The multiresonator is a set of cascaded spiral resonators designed to resonate at particular frequencies and create stop bands. The stop band resonances introduce magnitude attenuation and phase jumps to the transmitted interrogation signal at their resonant frequencies, which are

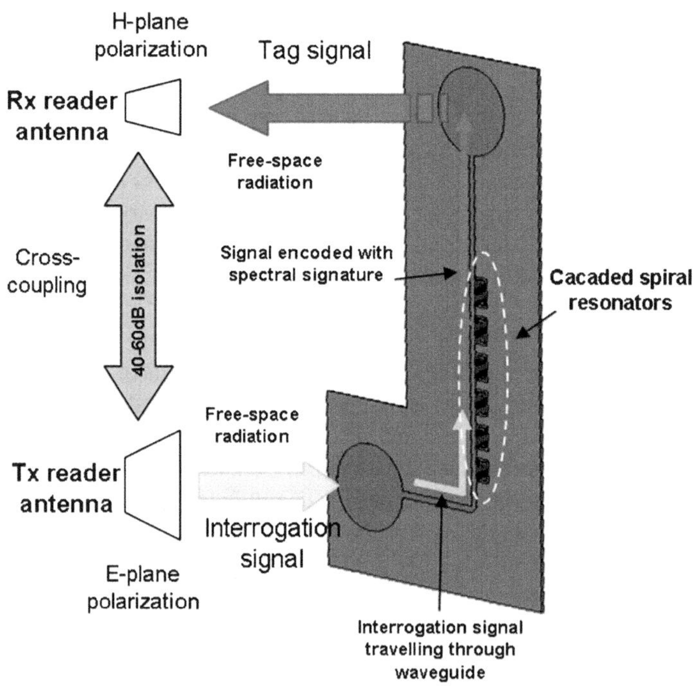

Fig. 5.1 Chipless RFID system signal flow diagram

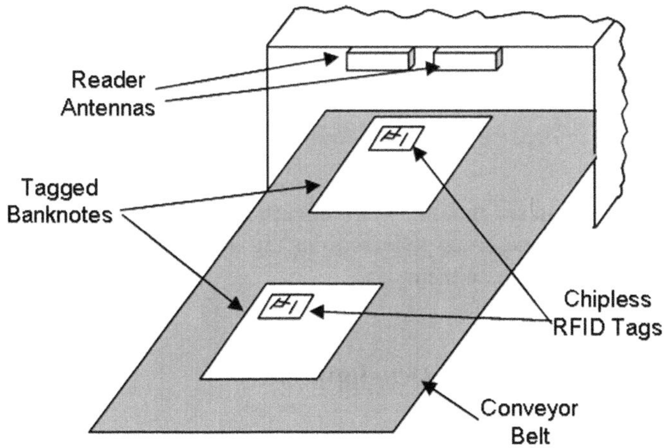

Fig. 5.2 Potential conveyor belt application for proposed chipless RFID system

detected as abrupt amplitude attenuations and phase jumps by the RFID reader. In order to provide isolation between the transmitting and receiving signal, the reader and tag antennas are cross-polarized. As a result, cross-talk between the transmitting and receiving antennas is minimized at the cost of introducing restrictions in tag positioning and orientation.

The proposed chipless RFID system is a preliminary design for a short range conveyor belt system, where the tagged items are tracked by moving through the interrogation zone of a fixed reader antenna system as shown in Fig. 5.2.

The expected power levels of the received signals from the chipless tags in an anechoic chamber (loss-less environment) in the antenna far-field region can be calculated using Friis free-space transmission formula [143]. The power density of the signal that reaches the chipless RFID tag in free-space is given by

$$S = \frac{P_t G_r}{4\pi r^2}, \tag{5.1}$$

where P_t is the transmitted power and G_r is the gain of the reader transmitting antenna and r is the separation distance between the tag and reader antennas. The power collected by the tag's antenna is defined as

$$P_a = SA_e = S\frac{\lambda^2}{4\pi}G_t, \tag{5.2}$$

where A_e is the effective area of the antenna defined by G_t, the tag's antenna gain, and wavelength λ. Hence, the signal received by the reader after interrogating the tag is defined as

$$P_{rx} = \frac{P_t G_t^2 G_r^2 \lambda^4 L(f)}{(4\pi r)^4}, \tag{5.3}$$

where $L(f)$ is the insertion loss of the tag's multiresonating circuit as a function of frequency f. The theoretical polarization loss factor p between two linearly polarized antennas with a misalignment angle θ can be expressed as in ref. [144]:

$$p = \cos^2 \theta. \tag{5.4}$$

From (5.4), we can see that the minimum efficiency (maximum isolation) is for $\theta = 90°$. Hence, we have cross-polarized our tag and reader antennas in order to minimize cross-talk between them.

5.3 Chipless RFID Tag Development

The tag design steps are shown in Fig. 5.3. The tag design starts by designing the tag monopole antennas and achieving the necessary return loss bandwidth and radiation pattern. Following the tag antenna design, design and optimization of the spiral resonators (multiresonating circuit) are performed. When the tag antenna and multiresonating circuit are optimized, they are integrated to form a complete chipless RFID tag, which is then tested in a wireless experimental setup inside an anechoic chamber (for theoretical verification) and in a laboratory (for robustness investigation).

In the following sections, the design and testing of a 6- and 35-bit chipless tag on a standard microstrip laminate and a 23-bit chipless tag on flexible thin laminate are presented.

5.4 Design

The tag operating principle based on spectral signatures and polarization diversity was presented in the preceding section. Following the design of spiral resonators and UWB monopole antennas, the integration of these two tag components to form a chipless RFID tag is performed. The integrated tag is designed in the following steps:

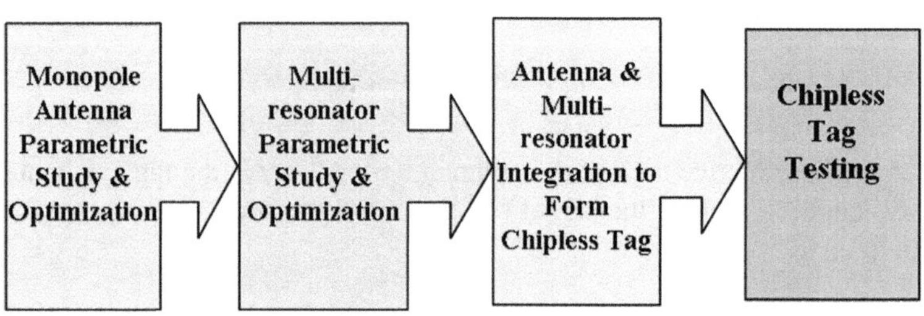

Fig. 5.3 Chipless tag design process

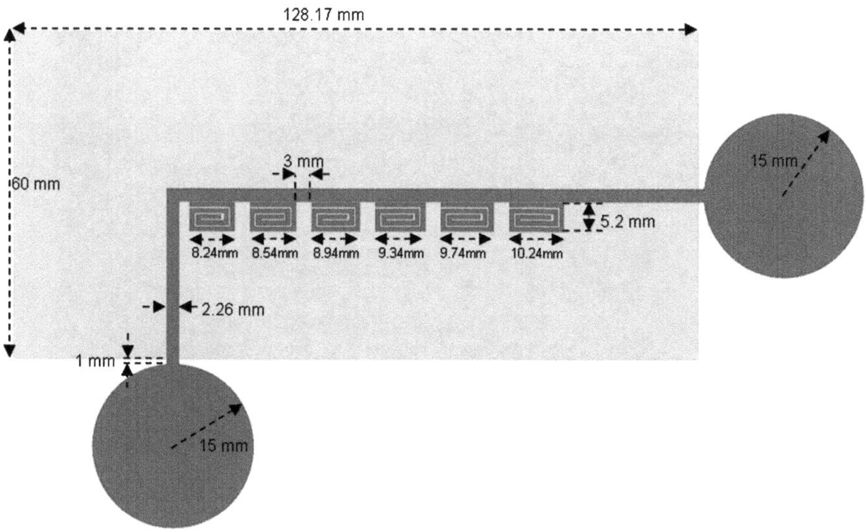

Fig. 5.4 Chipless tag layout with parameters on Taconic TLX-0 laminate ($\varepsilon r = 2.45$, $h = 0.787$ mm, $\tan\delta = 0.0019$)

- Initially, the UWB monopole antenna is simulated and measured. Good performance in the UWB spectrum in radiation patterns, return loss and gain is measured on Taconic TLX-0 PCB and TF-290 flexible laminate (Fig. 5.4).
- Six and 35-spiral resonators are designed cascaded on Taconic TLX-0 using microstrip technology between 2–2.5 and 3.1–7 GHz, respectively. The spirals are cascaded next to a 50 Ω microstrip line with 3 mm separation between spirals. Measured results showed good performance in both magnitude and phase of the spectrum.
- Twenty-three spiral resonators are designed cascaded on TF-290 flexible laminate using CPW technology. The spirals are cascaded within a 50 Ω strip line with 3 mm separation between spirals. Measured results showed good performance in both magnitude and phase of the spectrum.
- Finally, two UWB monopoles and cascaded spirals—multiresonator are integrated to form the chipless RFID tag. The monopole antennas are placed at each port of the 2-port multiresonator circuit to provide transmitting and receiving capabilities for the tag.

5.4.1 Six-Bit Chipless RFID Tag on PCB: Proof-of-Concept Tag

The layout of the tag is shown in Fig. 5.5. It consists of a vertically polarized UWB disc-loaded monopole receiving tag antenna, a multiresonating circuit and a horizontally polarized UWB transmitting tag antenna. In order to prove the concept of the novel chipless RFID tag, the operating frequency is from 2 to 2.5 GHz.

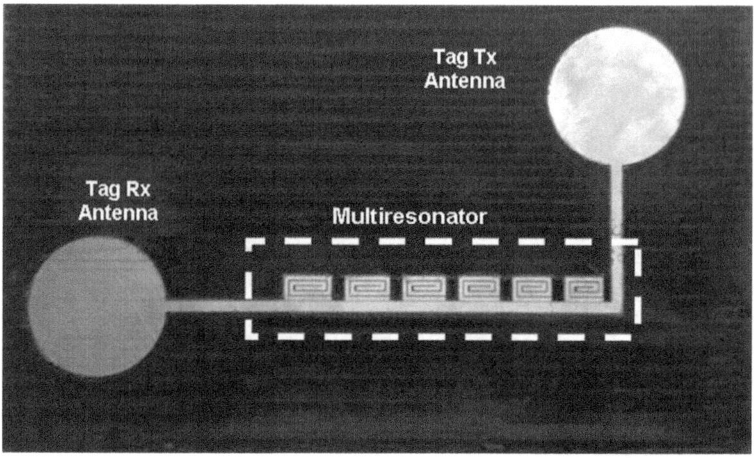

Fig. 5.5 Photograph of chipless RFID tag on Taconic TLX-0 laminate (εr=2.45, h=0.787 mm, tanδ=0.0019)

The photograph of the chipless RFID tag manufactured on Taconic TLX-0 (εr=2.45, h=0.787 mm, tanδ=0.0019) is shown in Fig. 5.5. The spiral resonators have different lengths in order to have different resonant frequencies as can be seen in Fig. 5.5. The spiral resonators are cascaded next to a 50 Ω microstrip line. The ground plane of the tag is etched/removed under the radiation patches of the monopole antennas.

5.4.2 UWB 35-Bit Chipless RFID Tag on PCB

The successful design and results of a 6-bit tag have motivated us to design a 35-bit chipless tag, which may find application for the Australian polymer banknotes and secured documents. In the previous section, we presented the proof-of-concept tag/system working in an unlicensed frequency band between 2 and 2.5 GHz with only 6 bits of data and significantly larger dimensions. In this section, we present a chipless RFID tag operating in the UWB spectrum (between 3.1 and 7 GHz) with a data capacity of 35 bits (capable of providing over 1.3 billion ID combinations). The number of bits in the UWB tag is limited by the second harmonic of the spiral resonators. The layout of the tag is shown in Fig. 5.6.

The dimensions of the chipless tag are limited mainly by the dimensions of the object being tagged, in this case an Australian polymer banknote. The 50$ Australian banknote is shown in Fig. 5.7 with its dimensions. Therefore, the tag must be designed to fit into the available area of 65 mm × 150 mm.

The photograph of the tag is shown in Fig. 5.8. The UWB chipless RFID tag consists of a vertically polarized UWB monopole receiving antenna, a 35-spiral

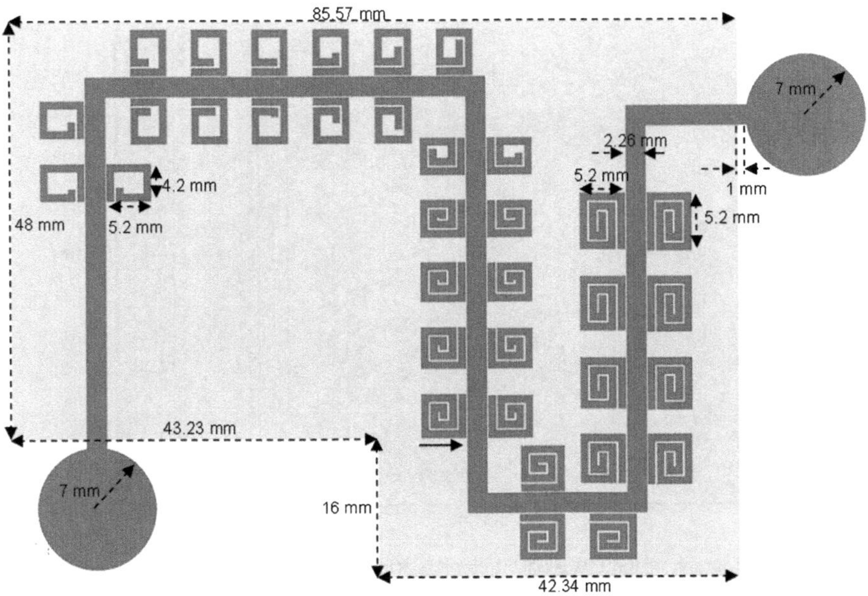

Fig. 5.6 Layout of integrated UWB 35-bit chipless tag with design parameters on Taconic TLX-0 laminate ($\varepsilon r = 2.45$, $h = 0.787$ mm, $\tan\delta = 0.0019$)

Fig. 5.7 Australian $50 banknote with dimensions

multiresonating circuit and a horizontally polarized UWB monopole transmitting antenna. The cross-polarized antennas minimize cross-talk between transmitting and receiving signal. In order to minimize the area of the tag, the spirals were placed on both sides of the meandered microstrip line.

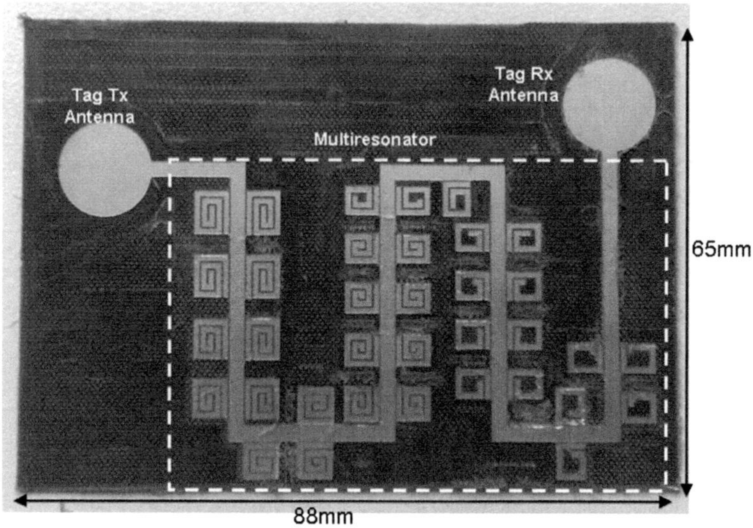

Fig. 5.8 Photograph of the UWB 35-bit chipless RFID tag with dimensions on Taconic TLX-0 laminate (εr = 2.45, h = 0.787 mm, tanδ = 0.0019)

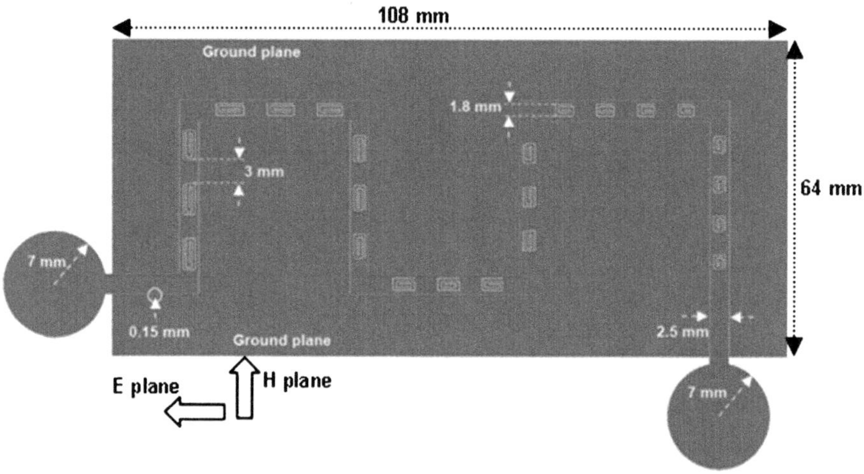

Fig. 5.9 Layout of integrated flexible CPW 23-bit chipless tag with design parameters on Taconic TF-290 laminate (εr = 2.9, h = 0.09 mm, tanδ = 0.0028)

5.4.3 UWB 23-Bit Chipless RFID Tag on Thin Flexible Laminate

The flexible chipless tag is designed on laminate Taconic TF-290 (εr = 2.9, h = 90 μm, tanδ = 0.0028) using ADS Momentum 2008. With this purpose, the antenna and multiresonators are designed individually. The layout of the chipless RFID tag with

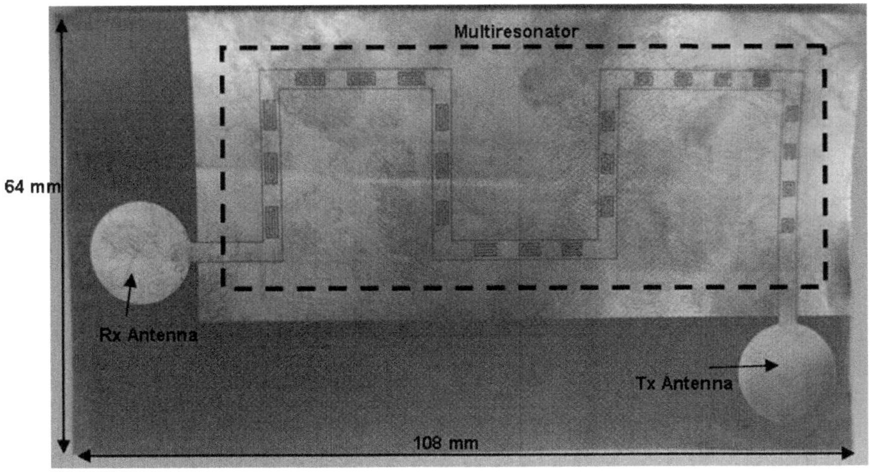

Fig. 5.10 Photograph of 23-bit chipless RFID tag on Taconic TF-290 ($\varepsilon r = 2.9$, $h = 90$ μm, $\tan\delta = 0.0028$)

design parameters printed on flexible TF-290 laminates is shown in Fig. 5.9. The tag is designed on CPW, making it single-sided.

The photograph of the tag is shown in Fig. 5.10. The tag encodes 23 bits of data between 5 and 10.7 GHz. The chipless tag is comprised of a vertically polarized UWB disc-loaded monopole receiving tag antenna, a multiresonating circuit and a horizontally polarized UWB transmitting tag antenna designed using CPW technology. The chipless tag is designed to fit the Australian banknote and its dimensions are 108 mm × 64 mm. The spirals were etched out with the spiral trace and separation between spiral traces being 0.2 mm. The 50 Ω CPW strip line was designed to be 2.5 mm with the gap separation from the ground plane being 0.15 mm. The spirals were etched in the strip line with a 3 mm separation between adjacent cascaded spirals.

Having presented all three designs of different chipless RFID tags in this section, the field trials of tags will be presented in the following section.

5.5 Field Trials

The field trials were performed in an anechoic chamber in order to validate the successful encoding of the tag and its detection at the reader end using Agilent's E8361A performance network analyser (PNA) as the reader. The chipless tag along with the reader antennas (which are UWB monopoles) has been mounted on plastic stands and placed in the anechoic chamber. The VNA was calibrated with the output power of the ports being −28 dBm. The block diagram and photograph of the experimental setup are shown in Fig. 5.11. Figure 5.11a shows that the reader antennas are cross-polarized in order to minimize cross-talk, and thus improving the isolation

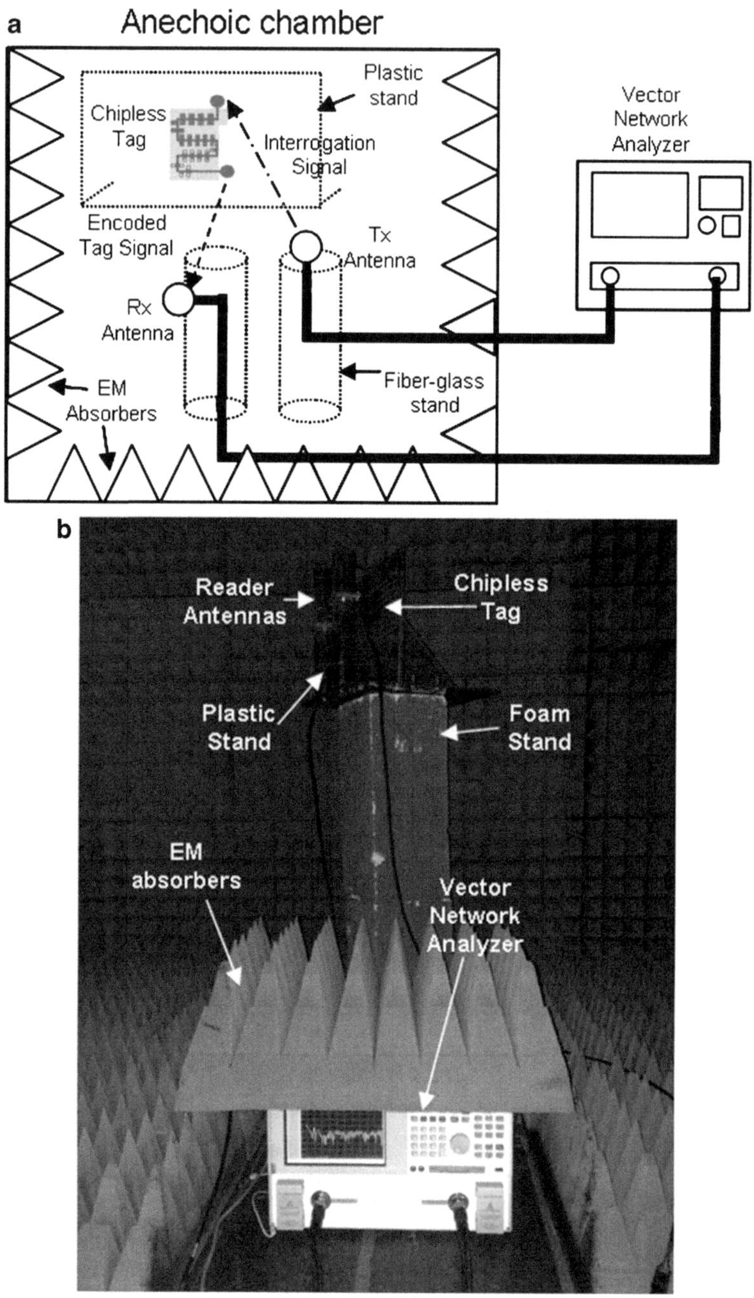

Fig. 5.11 Chipless RFID tag experiment (**a**) block diagram and (**b**) photograph

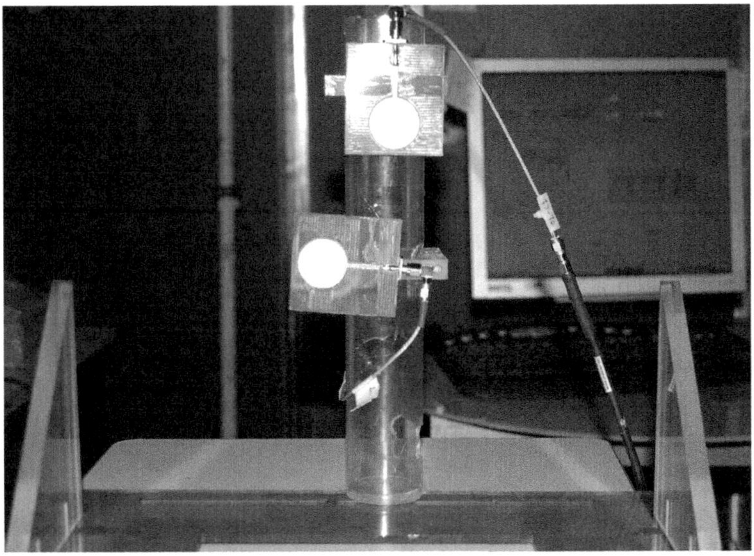

Fig. 5.12 Cross-polarized reader antennas mounted on a plastic stand

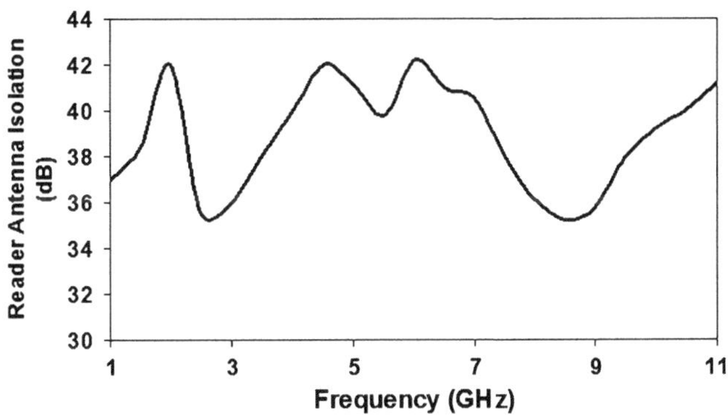

Fig. 5.13 Measured isolation between cross-polarized tag antennas

between the interrogation signal and the tag signal. Due to the monopole antennas small gain and omni-directional radiation pattern, a reading range of only a few centimetres was achieved. Greater reading ranges are achieved using high gain directional reader antennas and will be discussed in later chapters. Figures 5.12 and 5.13 show the cross-polarized reader antennas and the measured isolation of the UWB monopole reader antennas, which is greater than 35 dB, respectively.

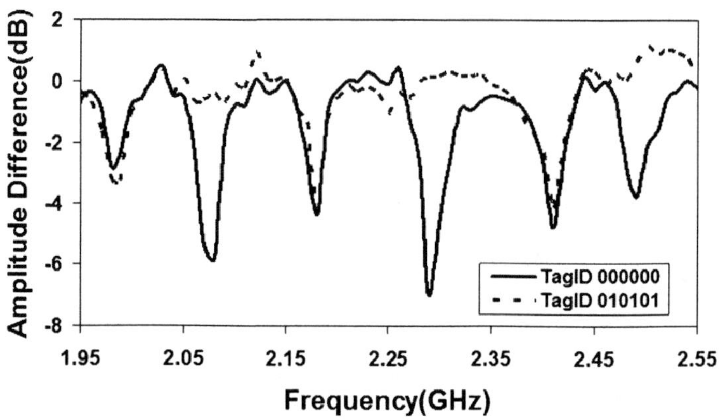

Fig. 5.14 Amplitude variations of the received tag signal at the reader end for chipless tags at 5 cm

The PNA sweeps the spectrum from the calibrated minimum frequency to the calibrated maximum frequency with equal output power of −28 dBm. The amplitude and phase information in the spectral signature are measured and assessed in such a way that the tag with no resonances (all 1's = no nulls and no phase jumps) is used as a reference. The following sections will provide the measured field trials of the proof-of-concept 6-bit tag, the UWB 35-bit tag on PCB and the UWB 23-bit tag on thin flexible laminate.

5.5.1 Six-Bit Chipless RFID Tag on PCB

The proof-of-concept tag encodes 6 bits of data from approximately 2 to 2.5 GHz. The measured results of the received data are presented in both magnitude and phase. The amplitude difference of the received power due to the tag frequency signatures is presented in Fig. 5.14. From the amplitude vs. frequency measurement, the distinct response of codes "000000" and "010101" are clearly visible. The worst case amplitude difference between 0 and 1 is more than 4 dB (at 2.5 GHz). The result reflects the robustness of encoding data of the proposed tag. This robustness is provided by the orthogonally polarized Tx and Rx antenna system for both the tag and the reader. Figure 5.15 shows the amplitude difference in respect to 111111 ID of the received signal after interrogating the same tag rotated by 180°, hence creating a polarization mismatch. The amplitude information is considered completely unreadable, and therefore useless.

Following the successful amplitude only measurement, the phase characteristics of the multiresonator-based tag is investigated. The phase information in the spectral signature was measured and compared with a reference tag with the 111111 ID

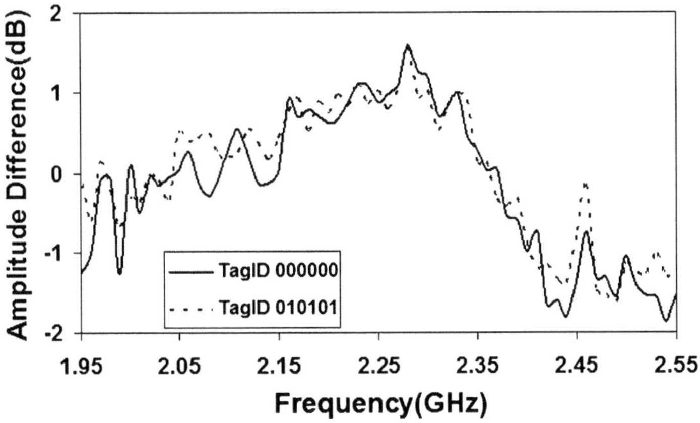

Fig. 5.15 Received signals by reader for chipless tags at 5 cm rotated by 180° (polarization mismatch between tag and reader antennas)

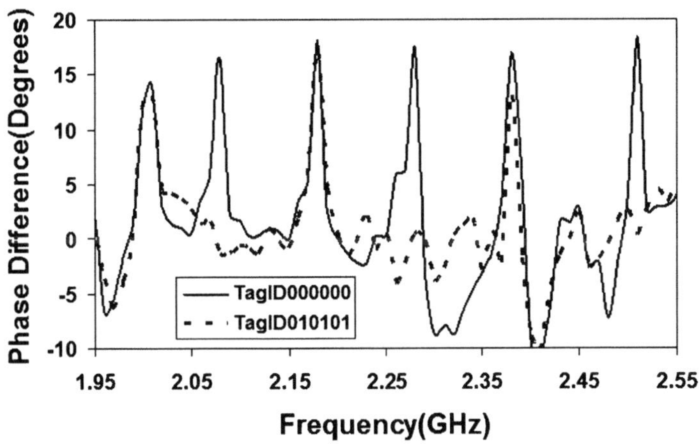

Fig. 5.16 Phase variations of the received tag signal at the reader end for chipless tags at 5 cm

used as a reference. The variation of the phase of the received power due to the tag's frequency signatures is presented in Fig. 5.16. From Fig. 5.16, it is clear that there is a significant phase difference of more than 15° between a logic "0" and logic "1". This is a significant finding in encoding bits in phase. A commercial phase detector can differentiate between single degrees of phase difference. Hence, no doubt, the phase-encoded tag would produce a more robust detection method compared to its amplitude encoding. This will be further demonstrated in the field trials of the

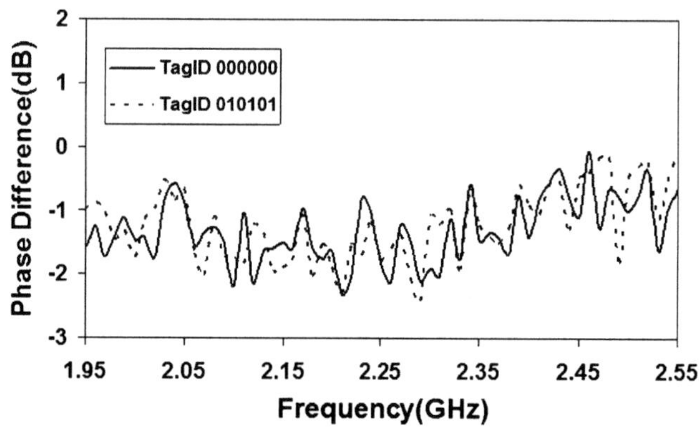

Fig. 5.17 Phase variations of the received tag signal at the reader end for chipless tags at 5 cm rotated by 180° (wrong alignment)

proposed chipless tag-reader system in the later chapters. As mentioned in the preceding section, the system is orientation sensitive due to the orthogonally polarized Tx and Rx antennas.

Figure 5.17 shows the phase information when the tag is rotated by 180° creating complete polarization mismatch with the reader antennas. The distinct phase difference between the two states completely disappeared making the tag unreadable. These orientation sensitive operations will make the proposed system most suitable for banknote authentication, where hundreds of notes pass through a conveyor belt system.

5.5.2 UWB 35-Bit Chipless RFID Tag on PCB

The UWB 35-bit chipless tag encodes 35-bits of data from 3.1 to 7 GHz. The 35-bit data is encoded using a 35-spiral multiresonator circuit on Taconic TLX-0 as shown in Fig. 5.18. The MSB spiral corresponds to the MSB data bit at 3.1 GHz as does the LSB spiral correspond to the LSB data bit at 7 GHz. The proof-of-concept phase of the 35-bit tag is conducted first in a wired condition using an Agilent 8361A PNA and then two orthogonally polarized monopole antennas are connected and tested in a wireless condition. The results of the comparative studies are presented in Figs. 5.19 and 5.20.

The amplitude difference of the received power due to the tag's spectral signature and the tag's multiresonator insertion loss are presented in Fig. 5.19. The variation of the received power's phase due to the tag's spectral signature is presented in Fig. 5.20. The multiresonator measurements are compared against the received tag's signal at the reader end, which corresponds to one another, hence confirming successful tag operation.

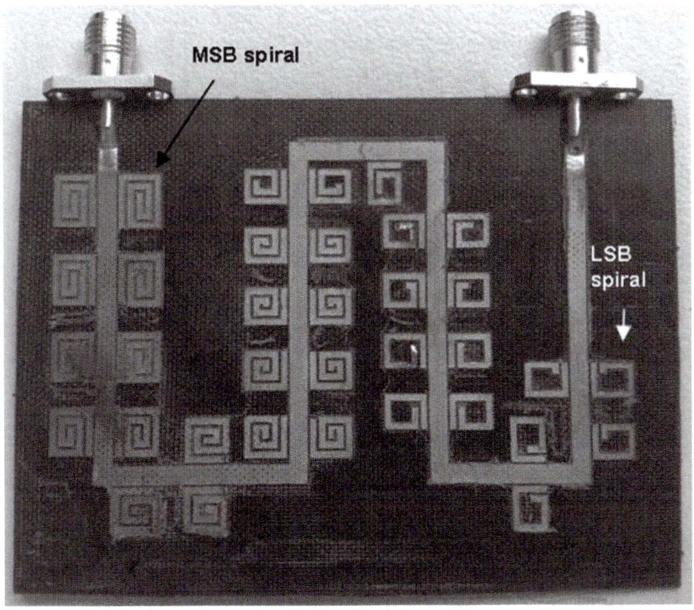

Fig. 5.18 Photograph of manufactured 35-bit chipless tag multiresonator on Taconic TLX-0

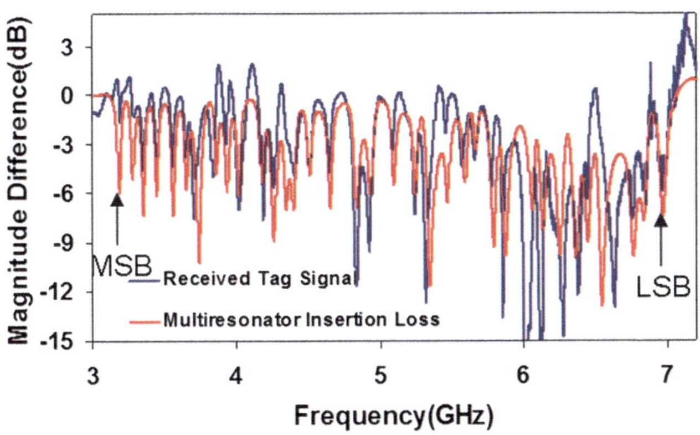

Fig. 5.19 Measured amplitude of tag spectral signature and multiresonating circuit

The tag field trials conclude the successful design, testing and wireless detection of the chipless RFID tag on PCB. However, our objective is to make the tag operate on a thin Australian polymer banknote. The following section presents the chipless tag designed on a thin flexible laminate with electrical properties similar to those of the polymer banknote substrate.

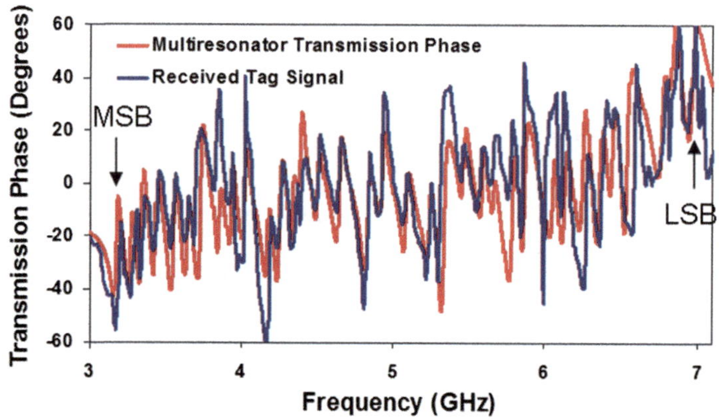

Fig. 5.20 Measured phase of tag spectral signature and multiresonating circuit

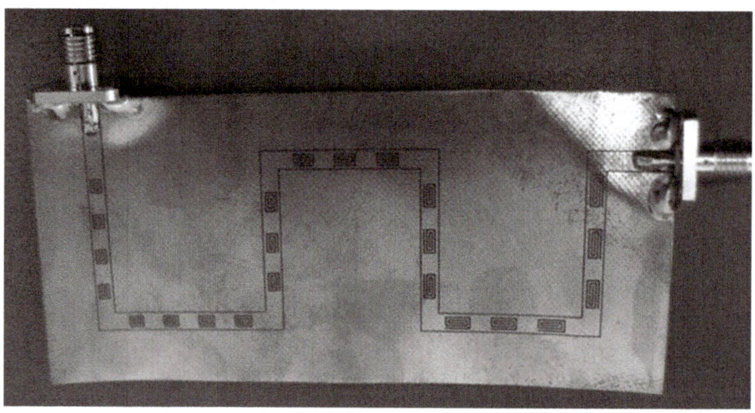

Fig. 5.21 Photograph of the 23-spiral multiresonating circuit on TF-290

5.5.3 UWB 23-Bit Chipless RFID Tag on Thin Flexible Laminate

The CPW-based UWB 23-bit chipless RFID tag encodes 23 bits of data from 5 to 10.7 GHz. The tag is printed on thin flexible laminate Taconic TF-290. The 23 bits of data are encoded using a 23-spiral multiresonating circuit shown in Fig. 5.21. The measured spectral signatures in both insertion loss vs. frequency and transmission phase vs. frequency of the 23-bit tag are shown in Figs. 5.22 and 5.23, respectively. From Figs. 5.22 and 5.23, it is clear that the 23 logic "0" bits are detected in the magnitude as magnitude nulls (or dips), while their phase signature is represented by 23 phase jumps. These measurements confirm the successful operation of the tag.

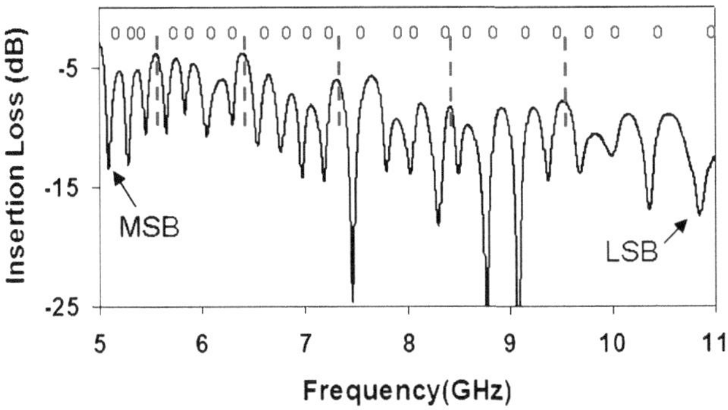

Fig. 5.22 Measured tag insertion loss of 23 bit tag ID “0x000000”

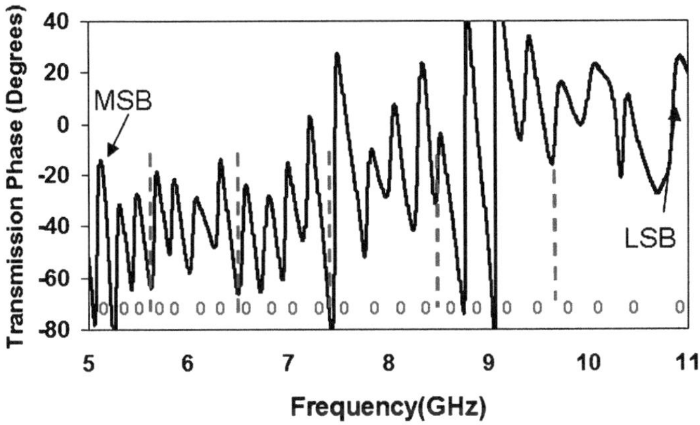

Fig. 5.23 Measured tag transmission phase of 23 bit tag ID “0x000000”

The successful detection of the 6- and 35-bit tags on PCB and the 23-bit tag on thin flexible substrate conclude the successful design and integration of the fully printable chipless RFID tags and prove the novel chipless tag concept.

5.6 Conclusion

In this chapter, the concept of multiresonator-based chipless RFID, integration of UWB monopoles and spiral resonators to form a chipless tag and tag field trials are presented. The theory of operation of the chipless RFID tag is explained for use in conveyor belt applications where a single tag is read at a time.

Three different chipless RFID tags were designed. The first tag that was designed was the proof-of-concept 6-bit tag on Taconic PCB substrate. The chipless tag operated between 2 and 2.5 GHz with 6 bits of data encoded by 6-spiral resonators.

The second tag was the UWB 35-bit tag on Taconic PCB as well. The UWB 35-bit tag operated from 3.1 to 7 GHz. The third tag was designed on thin flexible laminate Taconic TF-290 and encoded 23 bits of data from 5 to 10.7 GHz. All three chipless tags were presented comprehensively in terms of layout design and dimensions. The 35-bit tag on PCB and 23-bit tag on thin laminate were designed so that they can fit on the Australian banknote. Tagging of Australian banknotes for anticounterfeiting and identification applications was the primary motive for determining the size of the chipless tag.

The designed chipless tags were tested using an experimental setup placed inside an anechoic chamber. The tags were interrogated by an Agilent's vector network analyzer PNA 8361A. The reader antennas were UWB monopole antennas, which were cross-polarized. The VNA was used to sweep the desired frequency band and record the insertion loss recorded between the reader antennas. The successful detection of all three tags in both magnitude and phase of the spectral signature was confirmed. Thus, the concept of mulitresonator-based chipless RFID utilizing cross-polarized antennas was proven. Also, the use of phase data encoding and extraction was the first of its kind to be investigated and reported in spectral signature chipless RFID development. The use of phase encoding along with amplitude data encoding allows for comparison between the two for more robust and accurate readings.

The following chapters will concentrate on the design of the chipless RFID tag-reader circuit in order to complete the system design.

Chapter 6
Transceiver Design for Chipless RFID Tag Reader

Abstract This chapter focuses on the design of three generations of chipless RFID tag reader transceivers. The Gen-1 and Gen-2 transceiveres operate between 2 and 2.5 GHz and are designed to read the 6-bit proof-of-concept chipless tag. However, the Gen-1 reader only has amplitude detection capability, while the Gen-2 reader has amplitude and phase detection capabilities. Finally, the third transceiver is developed to operate in the UWB frequency range between 5 and 10.7 GHz so it can detect a UWB chipless tag. The specifications, design and measurements of the transmitter and receiver chains are presented along with the overall power consumptions of each transceiver.

6.1 Introduction

In the preceding chapters, the design of chipless RFID tags on PCB and thin flexible laminates is presented. The integration and successful field trials of the chipless RFID tags proved the concept of multiresonator-based chipless RFID tag. It was also presented earlier that the tag's main attractive features were it is low cost, robustness, printability and ease of data encoding. These advantages are the salient features of the chipless RFID tag, which relinquishes the requirement of silicon chips.

The successful design and testing of the novel chipless tag demands the development and design of a chipless tag reader. Firstly, the design of a proof-of-concept chipless tag reader which operates over 2–2.5 GHz only and reads a 6-bit tag is attempted. Two designs: first generation (Gen-1) reader, which decodes only the amplitude of the spectral signatures of the chipless tag, and second generation (Gen-2) reader, which decodes both the amplitude and phase of the chipless tag's spectral signatures. Finally, a chipless tag reader which operates over the UWB spectrum 5–10.7 GHz is presented. The UWB tag reader uses down-conversion blocks in the RF transceiver topology in order to process the received tag signal at an intermediate frequency (IF) band.

S. Preradovic and N.C. Karmakar, *Multiresonator-Based Chipless RFID: Barcode of the Future*, DOI 10.1007/978-1-4614-2095-8_6,

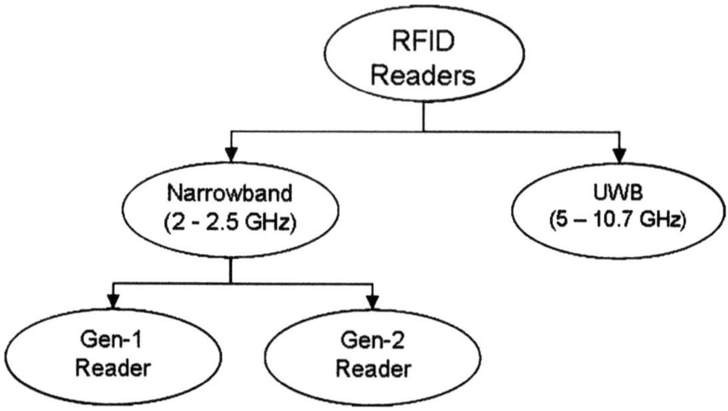

Fig. 6.1 Developed chipless tag RFID readers

As presented in Chap. 2, an RFID reader consists of three main parts: antenna(s), an RF transceiver and a digital/control section. The three RFID readers, which are presented in this chapter and shown in Fig. 6.1, differ mainly by the configurations of the RF transceivers mainly and their frequency of operation. The transceiver topologies that are presented in this chapter are using two antennas and dedicated transmit and receive RF paths to send interrogating CW signal to the tag and receive the encoded returned signals from the chipless tag. Hence, the chipless RFID system resembles a bi-static radar, which use two separately located Tx and Rx antennas. It is also the unique transceiver design for the chipless tag readers that differ them from conventional HF and UHF RFID readers that are found in the market today. The use of conventional readers is not possible, since the chipless tags are unique by their nature of spectral signature-based data-encoding technique, while conventional tags use time-domain data-encoding techniques.

This chapter is organized as follows: the operating principle of the proposed chipless RFID tag compared to the conventional reader is presented. Next, the design and specifications of RF transceiver which determine the frequency band of operation, the modulation technique, the power requirement, etc. are presented. The design and development of the readers are presented followed by the result of the reader. The transceivers are tested by means of wired tag interrogation conditions (without antennas) in order to confirm their successful operations.

6.2 Differences Between Chipped and Chipless Tag Readers

As mentioned in Chaps. 1 and 2, the only commercially successful and fully operational chipless RFID system is the SAW-based chipless RFID designed by RFSAW© [145]. A fully dedicated chipless SAW tag reader [146] was designed by the RFSAW engineers in order to accommodate the unconventional modulating and data-encoding properties of the SAW tag. Hence, it is imperative to design a reader that can read the multiresonator-based chipless tag by decoding data from their spectral signatures.

Although the use of conventional off-the-shelf RFID readers would be preferable, the new chipless RFID tags demand a fully new development of the reader from scratch. Three main differences between the developed chipless RFID tag reader and conventional off-the-shelf RFID reader are: (1) conventional RFID readers operate mostly at HF (13.56 MHz), UHF (915 MHz) and microwave (2.45 GHz) bands, while the chipless tag reader operates outside these bands; (2) conventional readers use amplitude shift keying (ASK) and binary phase shift keying (BPSK) time-domain-based demodulation techniques, while the presented reader is decoding the tag by sweeping the microwave frequency spectrum and acquiring the tag's spectral signature; and finally (3) the proposed reader can process the tag data even after the tag has been read and left the interrogation zone, while conventional readers require the tag to be in the reader's interrogation zone due to handshaking algorithms between tag and reader.

The difference in frequency of operation between the chipless tag reader and conventional RFID readers is more than obvious. As most chipped RFID systems operate in industrial, scientific and medical (ISM) bands [147–149], which have narrow bandwidths of a few KHz up to 83 MHz (2.45 GHz ISM band) due to fee-less license options, the chipless system proposed in this book operates in UWB region, which has a bandwidth greater than 500 MHz.

The most commonly used data-encoding techniques for conventional RFID tags are ASK [150] and PSK [151]. Hence, RFID readers designed for detecting these tags also use these two modulation schemes for data encoding and decoding [152]. This fundamentally means that conventional readers cannot detect and identify the chipless tag with its spectral signatures and development of a dedicated chipless RFID tag reader is necessary.

The communication between conventional RFID tags and readers is usually performed by using a handshaking algorithm between the two [153]. Another option is the use of tag polling by the reader, where each tag ID is stored in the reader's data base [154]. The reader transmits the tag ID's from its database one at a time and awaits for the tag with the polled ID to respond. Both of these options require the tag to be in the interrogation zone of the reader. The reader's interrogation zone is the area around the reader, where the tag can receive the reader's signal and retransmit a signal back to the reader, which can be detected [155]. In the case of a chipless RFID system, the tag is fully passive and hence cannot respond to the reader's polling, therefore no handshaking algorithms are possible or needed. Instead, the reader can perform processing of the tag's signal even when the tag has left the interrogation zone of the reader due to this matter.

The radio frequency transceiver is the heart of the proposed reader. The transmitter section generates the CW interrogation signal, and the receiver section received the echoes from the tag. Both sections are analogue circuits and set the link budget of the complete system. Therefore, it is vital to formulate the specifications, such as frequency of operation, transmit power level, interrogation signal type, receiver sensitivity, etc. In the following section, the specifications of the transceiver for the chipless tag reader are presented.

6.3 Transceiver Specifications for Chipless Tag Reader

The initial step of the design procedures of the chipless RFID tag reader is to formulate the specifications for a particular application. The conveyor belt application for tagging Australian polymer banknotes with the fully printable chipless RFID tags was presented in Chap. 5. The specifications of the tag and reader have been presented in Chap. 2. Hence, the specifications for the reader's transceiver need to fit the reader specifications presented in Chap. 2 and carry extra parameters, such as power level of the interrogation signal, transmitter to receiver leakage/isolation and receiver sensitivity. One of the main governing parameters for formulating the specifications are bandwidth, design frequency and reading range.

The bandwidth of the transceiver is defined as the frequency band over which the transceiver exhibits tolerable transmitter and receiver performances. The frequency of operation is set differently for different generations of readers. As for an example for the UWB reader, the transceiver is to operate within the UWB region (5–10.7 GHz). However, for Gen-1 and Gen-2, the transceivers are designed outside the UWB region for 2–2.5 GHz for less number of data bits and as the initial stage of the proof-of-concept exercise.

The reading range of the RFID reader is determined by the reader's transmitting power and the receiver's sensitivity in order to ensure successful tag interrogation and detection. Transmitting power limitations are dictated by standards, while the receiver's sensitivity is limited by bandwidth and minimum signal power.

Another important transceiver parameter which determines the transceiver architecture greatly is the type of interrogation that the reader is to perform. In this project, two specific options are explored:

1. Frequency sweep interrogation.
2. Pulse interrogation.

Frequency sweep interrogation is defined as sweeping a CW signal over an operating frequency band in a specific time interval (similar to a spectrum analyser). The advantages of this type of interrogation over pulsed interrogation are that the transmitter architecture is relatively simple to design and control [a voltage controlled oscillator (VCO) sweeps over a frequency band] and the receiver sensitivity is quite high. The disadvantages of this type of interrogation would be slower reading rates than those with a pulsed interrogation.

Pulse interrogation is defined as the interrogation of the chipless tag using a wideband pulse (or series of pulses) centred round the centre frequency of operation. The advantage of this type of interrogation is its high reading rates. The disadvantages are the need for ultra high speed switching components required for transmitter design, high speed ADC converters (above 1 GS/s) for receiver design and that the receiver detection is reduced [156] due to sinc ($\sin x/x$) function frequency distribution.

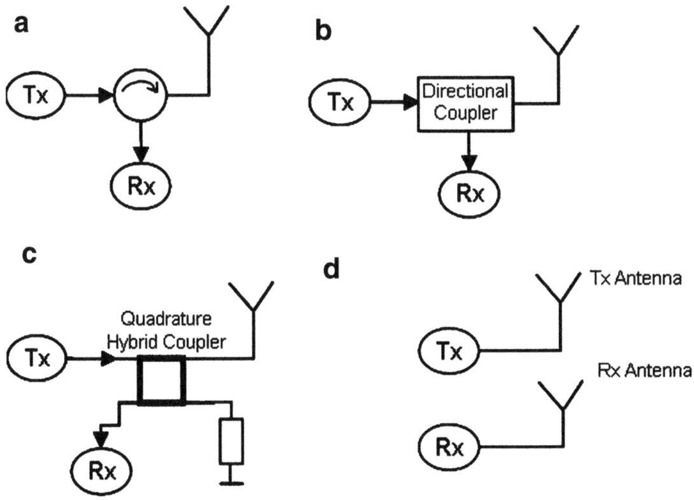

Fig. 6.2 Conventional RFID reader front end isolation architectures between Tx and Rx by using (**a**) circulator, (**b**) directional coupler, (**c**) hybrid coupler and (**d**) bi-static antennas

The isolation between the reader's transmitter and receiver is also important since the RFID reader must operate in full duplex mode. A duplexer cannot be used to remove the transmitter's leakage signal since the tag will respond at the same frequency as the transmitted signal [157]. Conventional RFID readers are designed with four transmitter isolation approaches to minimize the leakage [158–162]. The typically used architectures utilize isolation components, such as circulator (Fig. 6.2a), directional coupler (Fig. 6.2b), quadrature hybrid (Fig. 6.2c) and separate transmitter/receiver antenna, as shown in Fig. 6.2d.

The circulator, directional coupler and quadrature hybrid architectures are significantly dependant on the performance of the three aforementioned components. The use of two antennas does not increase the isolation significantly unless the reader antennas are linearly cross-polarized. The chipless tag reader transceiver designed in this project will have cross-polarized reader antennas for high Tx/Rx leakage cancelation.

As mentioned earlier, the design of the transceiver is conducted by designing two proof-of-concept transceivers (Gen-1 and Gen-2) operating from 2 to 2.5 GHz, which are used to detect the 6-bit proof-of-concept chipless tag. The third transceiver operates in the UWB region. Specifications for the Gen-1 and Gen-2 transceivers are given in Table 6.1, while the specifications for the UWB transceiver are given in Table 6.2.

The following section presents the design of the Gen-1, Gen-2 and UWB transceiver circuits.

Table 6.1 Specifications for the Gen-1 and Gen-2 transceivers

Electrical specifications	
Frequency of operation (GHz)	2–2.5
Transmitting power (dBm)	15
Interrogation signal type	Frequency sweep (CW)
Detection type	Amplitude only (Gen-1); amplitude and phase (Gen-2)
Tx/Rx isolation (dB)	60
Receiver sensitivity (dBm)	–35
Max. power consumption (Watts)	2
Commercial	
Cost	Less than AU $200 (a guide only)

Table 6.2 Specifications for UWB transceiver

Electrical specifications	
Frequency of operation (GHz)	Within 3.1–10.7
Transmitting power (dBm)	15
Interrogation signal type	Frequency sweep (CW)
Detection type	Amplitude and phase
Tx/Rx isolation (dB)	60
Receiver sensitivity (dBm)	–35
Max. power consumption (Watts)	10
Commercial	
Cost	Less than AU $2,000 (a guide only)

6.4 Design

In this section, the design of three types of transceivers for chipless RFID tag reader is presented. The transceiver circuits are designed to meet the specification presented in Sect. 6.3. Agilent's ADS Schematic and Momentum simulators were used to simulate the performances of the three transceivers before their fabrication and testing. The RF transceivers for the chipless RFID tag reader are designed and developed first followed by the digital circuit design, software programming and final system testing.

6.4.1 Gen-1 Transceiver

The Gen-1 transceiver is designed to operate between 2 and 2.5 GHz and detect the amplitude variations of the tag's spectral signature. In order to accomplish this, the receiver circuit utilizes a Schottky diode rectifier/detector circuit, which converts

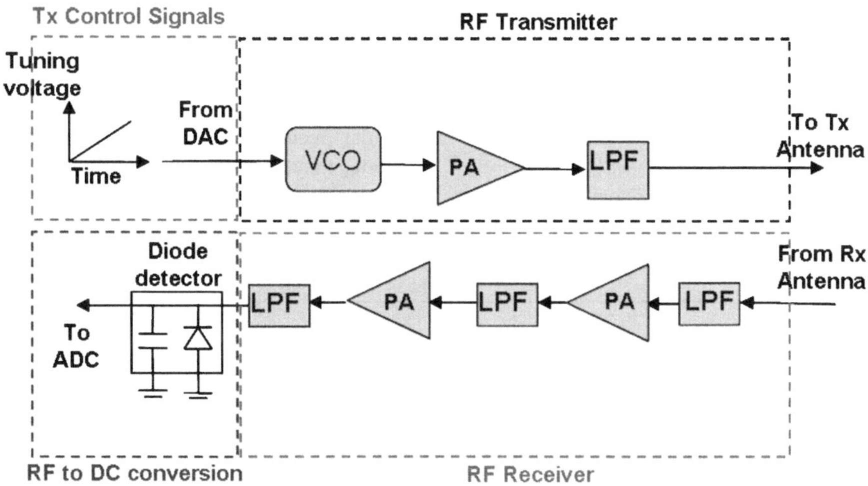

Fig. 6.3 Block diagram of Gen-1 transceiver

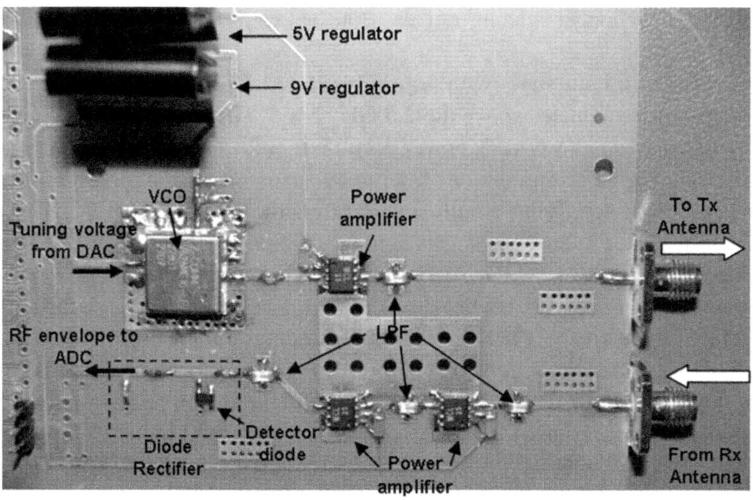

Fig. 6.4 Photograph of Gen-1 transceiver

the RF signal to an equivalent DC output. The DC signal is then sent to the reader's ADC for conversion to a digital signal. The block diagram and photograph of the Gen-1 transceiver are shown in Figs. 6.3 and 6.4, respectively.

The Gen-1 transceiver operates between 1.95 and 2.5 GHz and is consisted of a transmitter and receiver circuit. The RF components used for the design of the Gen-1 receiver are shown in Table 6.3. The control signals for the transmitter are sent from the DAC, which generates a tuning voltage for the VCO. The DAC generates a voltage

Table 6.3 Gen-1 transceiver RF component specifications

Component specifications	
V626ME10-LF VCO (Z-Communications)	Output frequency: 1.9–2.55 GHz; output power: 3–6 dBm
VNA-25 Power Amplifier (Mini-Circuits)	Operating frequency: 0.5–2.5 GHz; gain = 10 dB
LFCN-2500 Low Pass Filter (Mini-Circuits)	Operating frequency: 0–2.5 GHz
SMS7630-020 Schottky Diode (Skyworks)	Operating frequency: 0–10 GHz; voltage drop: 0.11 V
Commercial	
Cost	Less than AU $85

ramp from 0 to 14 V in order to sweep the output signal frequency generated by the VCO from 1.95 to 2.5 GHz. The VCO is a Z-Communications V626ME10-LF. The output power of the VCO is approximately between 3 and 6 dBm, which is below the required 15 dBm, so a Mini-Circuit VNA-25 power amplifier is added as a gain block. In order to filter out any higher order/frequency products, a Mini-Circuits LFCN-2500 low pass filter (LPF) is used. The amplified and filtered signal is thus sent to the broadband reader antenna, which can be a UWB monopole antenna for short reading ranges and a LPDA for longer reading ranges. Both antennas were comprehensively presented in Chap. 4.

The receiver circuit starts with an LFCN-2500 LPF, which is used to filter out any undesired frequencies above the 2.5 GHz band. The signal is further on amplified by two cascaded VNA-25 power amplifiers with low pass filtering stages in between. The signal is amplified in order to be an appropriate signal strength/power so that the diode-rectifying circuit may convert the RF power to DC. The diode rectifier consists of a shunt Skyworks Schottky diode and series 56 pf DC block and load capacitors and a 47 nH RF choke inductor. The circuit of the diode rectifier and Schottky diode model are made in ADS Analogue Schematic and are shown in Fig. 6.5a, b, respectively. The Schottky diode model is obtained from Skyworks Inc. The diode rectifier circuit acts as an RF envelope detector, which can be used to detect the dips in the RF power due to the chipless tag stop band resonances. The Schottky diode model takes into account the capacitive and inductive properties of the SOT-143 packaging of the SMS7630-020 diode. The SOT-143 has four ports/pins because there were two diodes in the SOT-143 packaging.

In the following section, the design of the Gen-2 RFID reader transceiver is discussed.

6.4.2 *Gen-2 Transceiver*

The Gen-2 transceiver is designed to operate between 2 and 2.5 GHz and detect both the amplitude and phase variation of the tag's spectral signature. In order to accomplish this, the receiver circuit utilizes an AD8302 gain/phase detector circuit

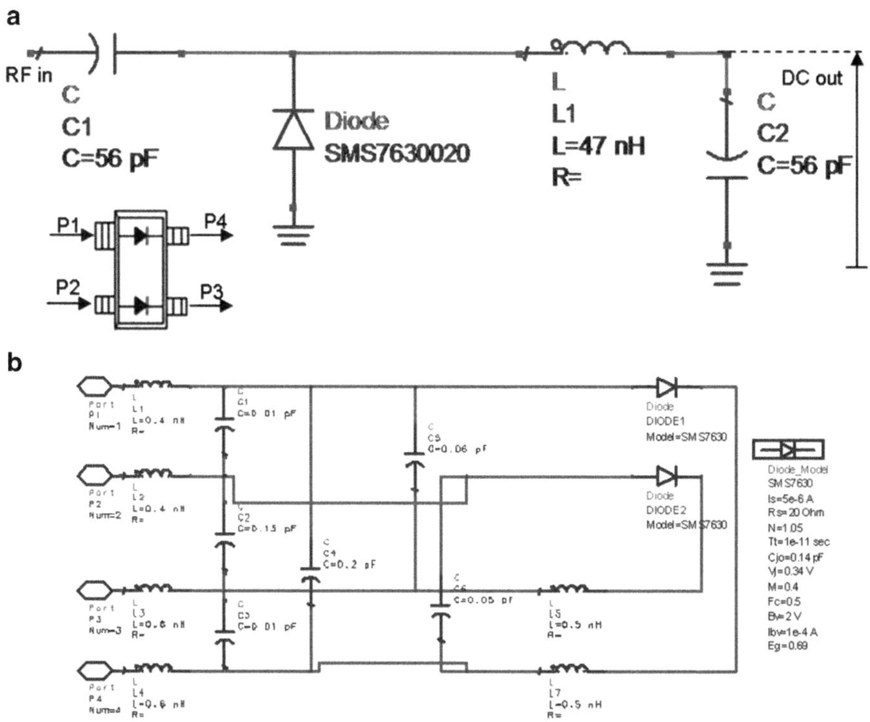

Fig. 6.5 Diode rectifier circuit (**a**) and diode model (**b**) designed using ADS 2008A

which generates a DC voltage output corresponding to the amplitude and phase difference between the reference RF signal and the received tag signal. The block diagram and photograph of the Gen-2 transceiver are shown in Figs. 6.6 and 6.7, respectively.

The RF components used for the design of the Gen-2 receiver are shown in Table 6.4. The transceiver consists of a transmitter and a receiver circuits. The transmitter circuit is controlled in the same way as the Gen-1 transceiver. The difference between the Gen-1 and Gen-2 transceiver is the power divider circuit, which provides a reference signal for the receiver circuit. The use of a single VCO means that the transceiver has a coherent architecture. The receiver circuit is the same as the Gen-1 except for the AD8302 gain/phase detector instead of the RF diode detector used in the Gen-1. The gain/phase detector is used to detect the variations in magnitude and phase of the received tag signal when compared to the reference signal supplied from the VCO. The AD8302 is multiplexed with a SN74LV4066A 8:1 multiplexer due to the use of a signal ADC on the digital control section of the RFID reader.

The Gen-2 receiver is advanced in comparison to the Gen-1 receiver since it has both amplitude and phase spectral signature detection and decoding capabilities. Both amplitude and phase detection and decoding capabilities of the Gen-2 transceiver yield robust reading of the chipless RFID tag. This advantageous feature of the Gen-2

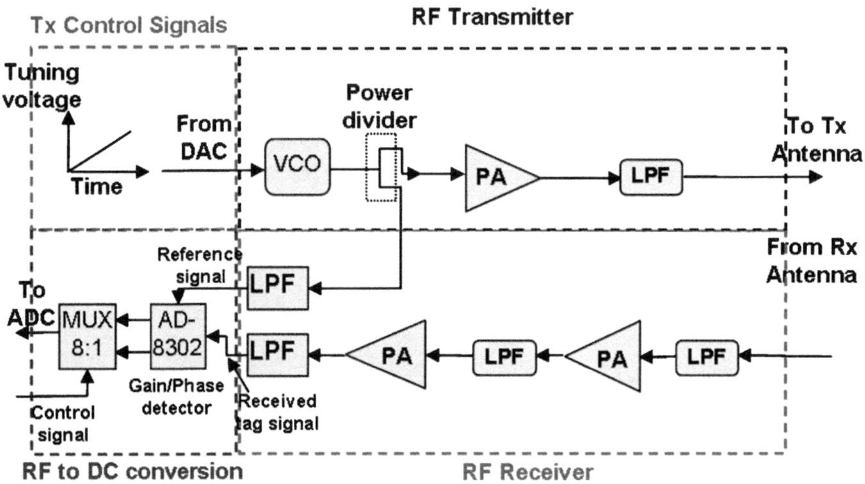

Fig. 6.6 Block diagram of Gen-2 transceiver

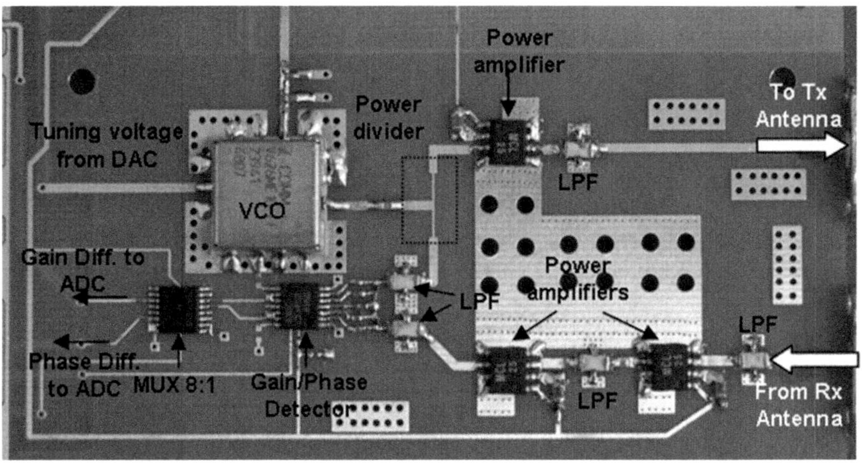

Fig. 6.7 Photograph of Gen-2 transceiver

transceiver will be presented in the result section (Sect. 6.5.2). With the design of the Gen-2 transceiver the proof-of-concept RFID transceiver circuit is complete. The next step in transceiver design for dedicated chipless tag RFID reader is the design of a UWB transceiver, which has the capability to work above 2.5 GHz.

Table 6.4 Gen-2 transceiver RF component specifications

Component specifications	
V626ME10-LF VCO (Z-Communications)	Output frequency: 1.9–2.55 GHz; output power: 3–6 dBm
VNA-25 Power Amplifier (Mini-Circuits)	Operating frequency: 0.5–2.5 GHz; gain = 10 dB
LFCN-2500 Low Pass Filter (Mini-Circuits)	Operating frequency: 0–2.5 GHz
AD8302 Gain/Phase Detector Analog Devices	Operating frequency: 0.3–2.7 GHz; voltage output: 0–1.8 V
SN74LV4066A 8-1 Multiplexer (Texas Instruments)	Operating frequency: DC; voltage output: 0–5 V
Commercial	
Cost	AU $110

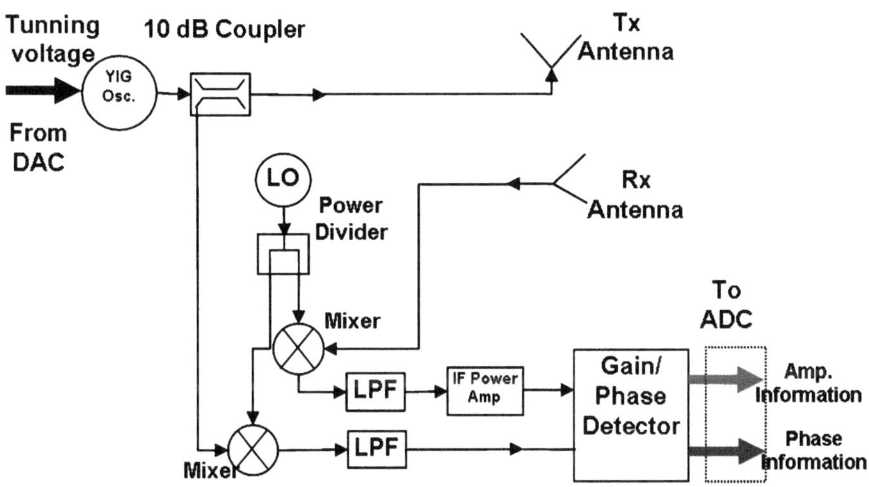

Fig. 6.8 Block diagram of UWB transceiver

6.4.3 *UWB Transceiver*

The UWB transceiver is designed to operate with 3.1 and 10.7 GHz and detect the amplitude and phase of the tag's spectral signature. Therefore, it can be stated that the UWB transceiver circuit is an extension of the Gen-2 transceiver. The UWB receiver circuit utilizes the AD8302 gain/phase detector circuit, which generates a DC voltage output corresponding to the amplitude and phase differences between the two RF signals—the reference from the transmitter and the received signal from the tag. The block diagram and photograph of the UWB transceiver are shown in Figs. 6.8 and 6.9, respectively.

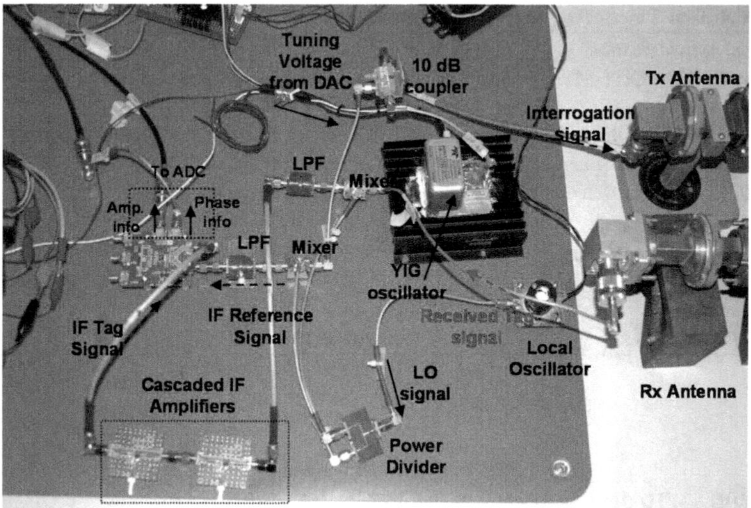

Fig. 6.9 Photograph of UWB transceiver

As can be seen in Fig. 6.8, the UWB transceiver consists of a transmitter and receiver path. The transmitter consists of a powerful Teledyne YIG oscillator, which operates from 2 to 6 GHz and generates an interrogation signal of constant 15 dBm power to the Tx antenna. Therefore, the power amplifiers are not needed in the transmitter path. The highest operating frequency of the UWB transceiver is limited by the highest operating frequency of the YIG oscillator, which can be extended by either upgrading the currently used oscillator or by adding a YIG oscillator working from 6 to 10.7 GHz. Table 6.5 shows the components used for the UWB transceiver design.

From Figs. 6.8 and 6.9, it is clear that the UWB transceiver utilizes a downconverting stage at the receiver, which consists of a pair of ZX05-14+ mixers. The UWB transceiver requires a pair of mixers since it has two IF channels fed into the AD8302 gain/phase detector unit: IF reference signal and IF tag signal. The IF signals cannot be above 2.7 GHz because the gain/phase detector can work only up to 2.7 GHz. The IF reference signal is obtained by downconverting the RF signal separated from the YIG oscillator by a Narda 4014C-10 10 dB coupler. The IF tag signal is obtained by downconverting the received tag signal. The local oscillator (LO) is a Richardson RVC6000 VCO, which operates from 4 to 8 GHz. After the down-conversion stage, the IF tag signal is filtered and amplified since the free-space loss attenuates the tag signal significantly. The IF amplifier circuits are Mimix broadband GMX7001 gain blocks, which give approximately 22 dB gain at 2 GHz each (cascaded over 40 dB gain). As discussed earlier, the gain/phase detector AD8302 provides a 0–1.8 V analogue DC voltage output at it is two output ports. The first port corresponds to the amplitude difference between the two IF signals, and the second port corresponds to the phase difference between the two IF signals.

Table 6.5 UWB transceiver RF component specifications

Component specifications	
YIG Oscillator (Teledyne)	Output frequency: 2–6 GHz; output power: 15 dBm
RVC6000 VCO (Richardson Electronics)	Operating frequency: 4–8 GHz; output power: 5 dBm
4014C-10 10 dB Coupler (Narda)	Operating frequency: 3–10 GHz
ZX05-14+ Mixer (Mini-Circuits)	Operating frequency: 4–9 GHz; attenuation: 24 dB
LFCN-2500 Low Pass Filter (Mini-Circuits)	Operating frequency: 0–2.5 GHz
AD8302 Gain/Phase Detector Analog Devices	Operating frequency: 0.3–2.7 GHz; voltage output: 0–1.8 V
Gain Blocks GMX7001 (Mimix Broadband)	Operating frequency: 0–6 GHz; gain: 22–14 dB
Commercial	
Cost	AU $3,800

Since only the tag's IF signal changes depending on the tag's spectral signature, the use of the reference signal decoupled from the transmitter path by a 10 dB coupler enables the detection of amplitude and phase variations in the tag's spectral signature. The analogue DC outputs of the gain/phase detector are passed to the digital section for further processing.

The UWB transceiver circuit is the ultimate design for interrogating the chipless tag, receiving the tag's response signal and processing it for decoding the spectral signature of the chipless tag in both amplitude and phase. This is due to the fact that it has used most stable transmitter and highly sensitive receiver, which comprises expensive YIG oscillator, couplers, mixers, filters and the gain/phase detector. Therefore, it is expected that the most robust results are obtained from the UWB transceiver. The following section provides the results of the three transceiver topologies.

6.5 Results

The measurement results of the three transceivers designed for the chipless tag RFID reader are presented in this section. The transceivers are tested individually and in a wired scenario. The wireless field trials of the integrated tag reader system will be presented in the next chapter. The wired testings are performed in order to validate the operation of the RF transceiver circuits only.

The block diagram of the experimental setup used for testing the transceivers for tag spectral signature detection is shown in Fig. 6.10. The RF output of the transceiver circuit is connected to the multiresonator circuit through a pair of attenuators at each port in order to minimize the RF power going into the receiver of the transceiver. The attenuators are necessary in order to prevent the transmitter output power sending the receiver into deep saturation or even destroying it. The pair of attenuators provides approximately 36 dB of attenuation.

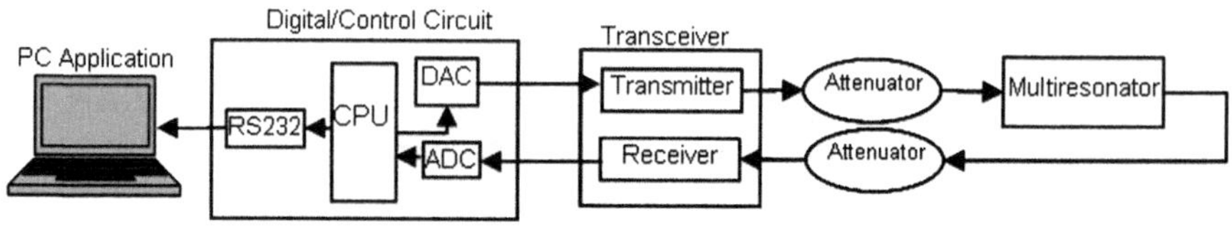

Fig. 6.10 Block diagram of transceiver testing experimental setup for tag detection

The digital/control circuit is designed to control the transmitting frequency of the transmitter circuit and digitize the acquired data from the received tag signal. The digital circuit CPU is an Atmel AT89C52 microprocessor. The CPU sends an 8-bit digital word to the DAC, which is converted to an analogue-tuning voltage for the VCO. The received tag signal is converted to an 8-bit digital word by the ADC. The CPU performs data decoding or can transmit data through the RS-232 serial interface to a software application installed on a PC.

To verify that the developed RF transceiver sections work correctly, an Agilent performance network analyser (PNA) E8361A is used. The two sets of data are compared. Upon obtaining the satisfactory agreement between the two results, the above procedure is implemented to digitize the analogue data from the transceiver circuit.

6.5.1 Gen-1 Transceiver

The following characteristic parameters of the Gen-1 transceiver are investigated and their effects on the performance of tag detection are recorded:

- Transmitter output power.
- Transmitting signal higher order harmonic distortion.
- Isolation between transmitter and receiver.
- Receiver sensitivity.

The measured transmitter output power is shown in Fig. 6.11. The RF output power is measured using Agilent's E8360 Spectrum Analyzer. The tuning voltage sent to the Gen-1 transmitter VCO is presented in Fig. 6.11 and shows that the frequency control is linear. Figure 6.11 shows that the transmitter output varies from 17.4 dBm at 2.1 GHz to 16.6 dBm at 2.5 GHz. The signal output drops at higher frequencies since the gain of the power amplifiers used drops as the signal frequency approaches 2.5 GHz. The measured output power of the transmitter VCO varies between 2 dBm (at 2 GHz) and 4 dBm (at 2.5 GHz) is shown in Fig. 6.12. The gain of the VNA-25 power amplifier drops by 3 dB between 2 and 2.5 GHz as shown in Fig. 6.13. Therefore, from Figs. 6.12 and 6.13, it is clear that the output power of the transmitter will drop by approximately 1 dB between 2 and 2.5 GHz. The measured return loss and insertion loss of the LPF are shown in Fig. 6.14. From

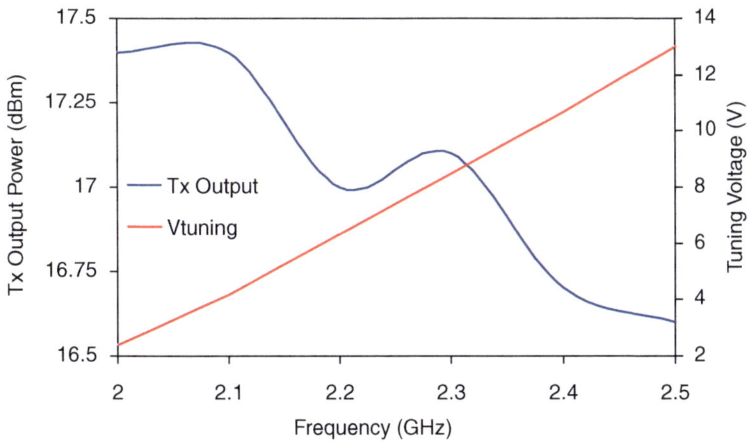

Fig. 6.11 Measured Gen-1 transmitter output power and VCO-tuning voltage

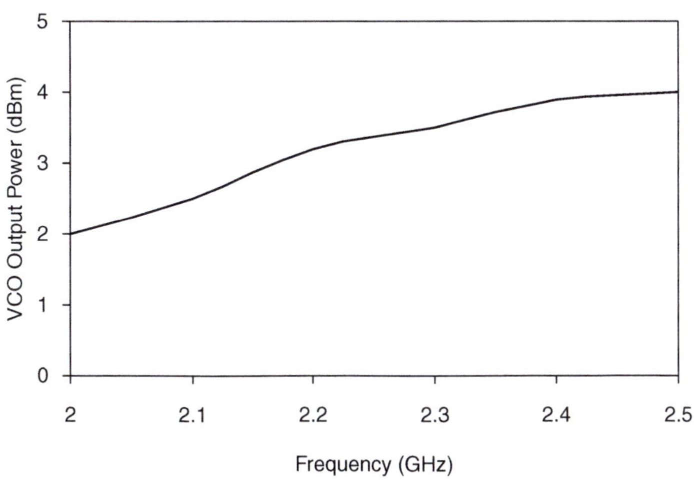

Fig. 6.12 Measured Gen-1 transmitter VCO output

Fig. 6.14, it is clear that the LPF exhibits less than 1 dB insertion loss and greater than 10 dB return loss below 2.5 GHz.

Another important characteristic of the Gen-1 transmitter is the presence of any higher order distortion products. Figures 6.15–6.17 show the measured spectrum of the transmitter output from 0 to 10 GHz when the transmitter output signal frequency is set to 2, 2.25 and 2.5 GHz, respectively. From Figs. 6.14–6.16, it is clear that the spectrum is "clean" of any higher frequency distortions due to the filtering section designed in the transmitter part (Fig. 6.18) (presented in Sect. 6.4.1).

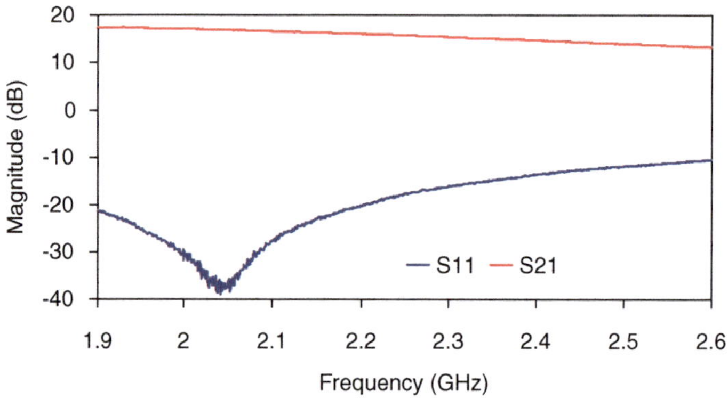

Fig. 6.13 Measured return loss (S11) and gain (S21) of VNA-25 power amplifier

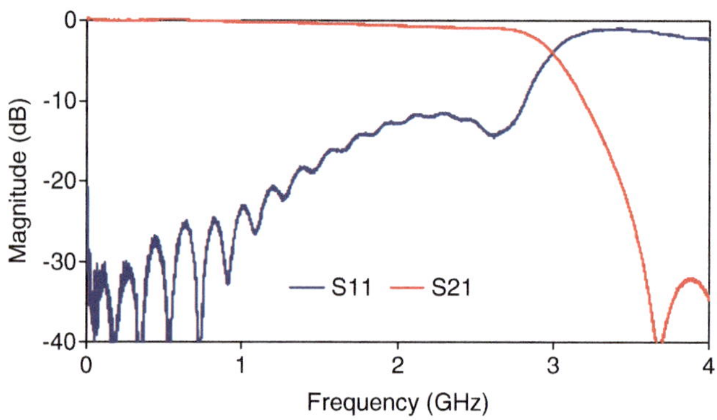

Fig. 6.14 Measured Gen-1 transmitter output power and VCO-tuning voltage

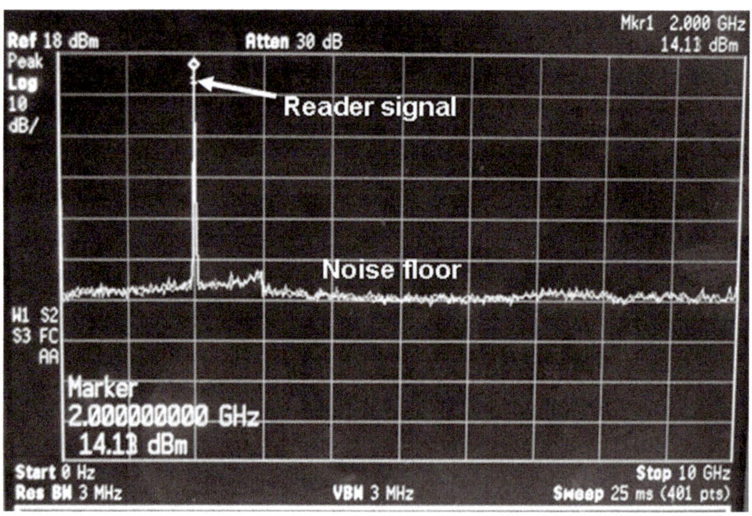

Fig. 6.15 Measured Gen-1 transmitter output spectrum from 0 to 10 GHz for signal output at 2 GHz

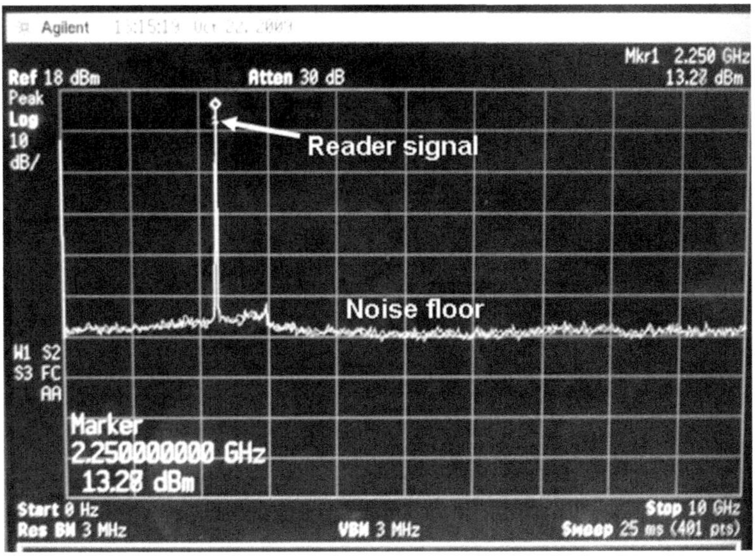

Fig. 6.16 Measured Gen-1 transmitter output spectrum from 0 to 10 GHz for signal output at 2.25 GHz

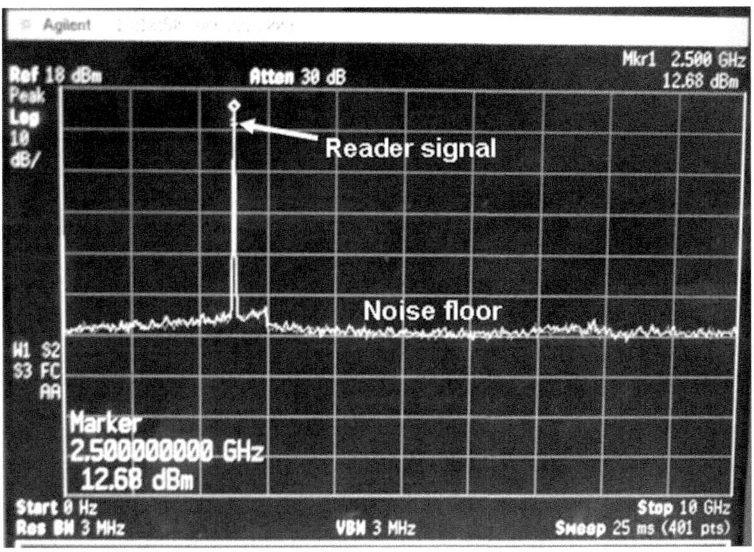

Fig. 6.17 Measured Gen-1 transmitter output spectrum from 0 to 10 GHz for signal output at 2.5 GHz

The isolation between the transmitter and receiver circuit determines the amount of power that "leaks" from the transmitter path to the receiver. The isolation between the transmitter and receiver was measured by terminating the transceiver's Tx and Rx ports with 50 Ω loads leaving the receiver with no input signal except the leakage signal through the ground. The power of the signal coming out of the receiver

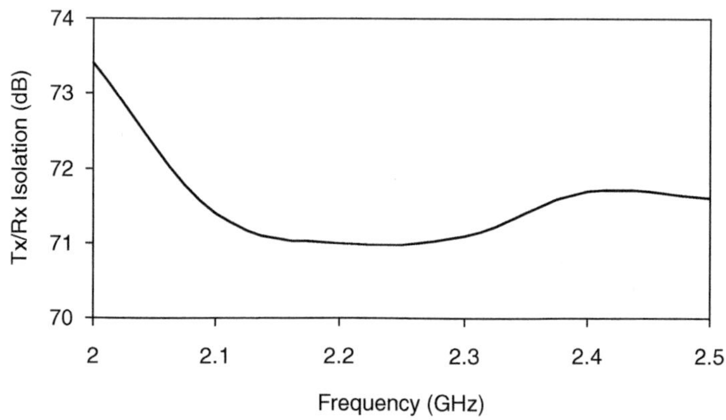

Fig. 6.18 Measured Gen-1 transmitter/receiver isolation

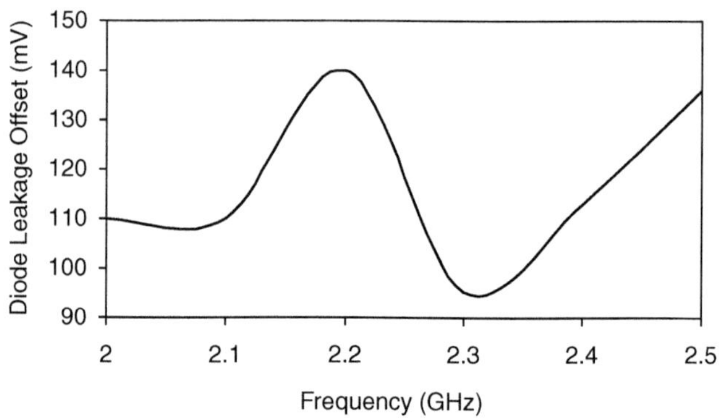

Fig. 6.19 Measured Gen-1 diode detector leakage offset

was measured and used to calculate the isolation between the transmitter and receiver path. Figure 6.18 shows the measured isolation above 71 dB.

The RF signal leakage from the transmitter to the receiver creates a DC offset at the output of the diode rectifier. This offset needs to be measured in order to recalibrate the receiver. The measured diode leakage offset is presented in Fig. 6.19. The nonlinear behaviour of the leakage offset is mostly due to the nonlinear diode detector rectification since the diode detector is a nonlinear device itself.

The receiver sensitivity was tested using the experimental setup shown in Fig. 6.20. Agilent's E8257D signal generator was used to determine the minimal signal power levels that can be detected between 2 and 2.5 GHz. The diode detector output was measured using Tektronix Oscilloscope with a 50 Ω coaxial probe. The receiver sensitivity determines the minimal received signal input power that can be successfully received by the Gen-1 transceiver. Figure 6.21 shows the measured Gen-1 receiver sensitivity. From Fig. 6.21, it is clear that the receiver sensitivity is

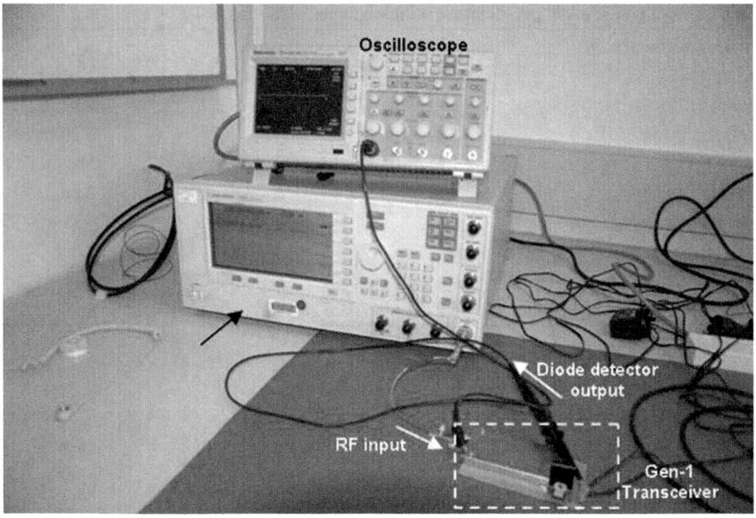

Fig. 6.20 Experimental setup for measuring receiver sensitivity

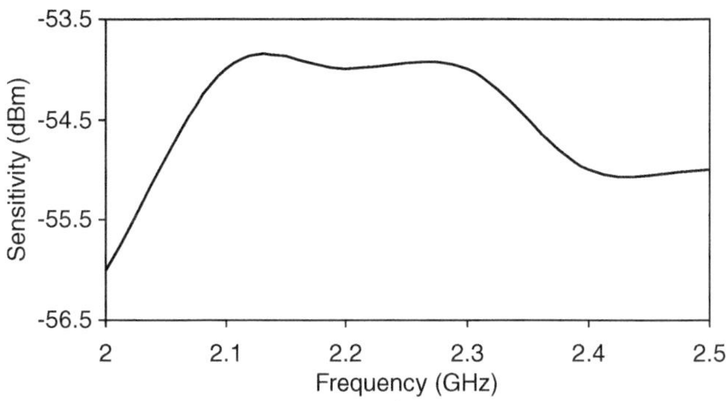

Fig. 6.21 Measured Gen-1 receiver sensitivity

below −53.6 dBm. The receiver sensitivity is mostly determined by the leakage signal from the transmitter. If the received tag signal after amplification is smaller than the leakage signal at the input of the diode detector, it is amplitude variation will not be noticed at the output of the diode detector. Further improvement of the receiver sensitivity is possible by increasing the isolation between the transmitter and receiver.

The Gen-1 transceiver characteristics are summarized in Table 6.6 along with the minimum specifications set in the design section. From Table 6.6, it is clear that the Gen-1 transceiver has met all of the desired requirements.

Table 6.6 Gen-1 transceiver pre-design and tested specifications

Electrical specifications		
Specifications	Pre-design	Tested
Frequency of operation (GHz)	2–2.5	1.98–2.5
Transmitting power (dBm)	15	Above 16.5
Interrogation signal type	Frequency sweep	Frequency sweep
Tx/Rx isolation (dB)	60	Above 71
Receiver sensitivity (dBm)	–35	–53 to –57
Max. power consumption (Watts)	2	1.5
Commercial		
Cost	Less than AU $200 (a guide only)	$85

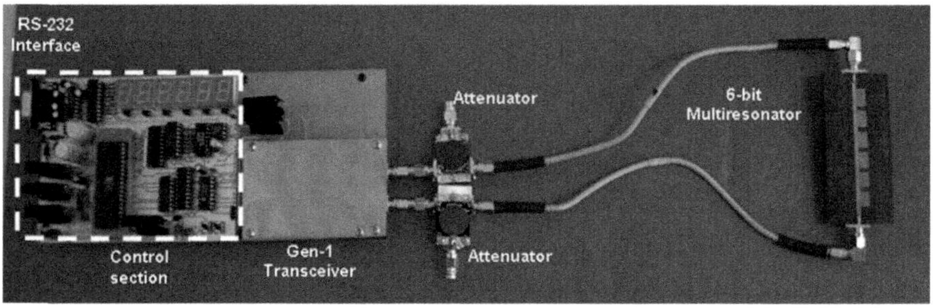

Fig. 6.22 Photograph of Gen-1 testing experimental setup for wired 6-bit tag detection

The Gen-1 transceiver is tested by connecting the tag multiresonator (without antennas) to the Tx and Rx ports as presented in Fig. 6.10. The Gen-1 transceiver is designed to detect only the amplitude of the spectral signature using a diode detector at the receiver end. The experimental setup of the transceiver integrated with the control section is shown in Fig. 6.22.

The transceiver transmits a linearly swept frequency variant CW signal from 1.98 to 2.5 GHz. The signal is received after passing through the 5-bit multiresonating circuit. An Agilent PNA E8361A network analyser is used to measure the spectral signature (insertion loss) of the 5-bit multiresonator. Figure 6.23 shows the measured spectral signature obtained from the PNA against the detected calibrated spectral signature using the Gen-1 transceiver. From Fig. 6.23, it is clear that the first three resonances of the multiresonator exhibit larger amplitude "dips" (nulls) since the gain of the VNA-25 power amplifier and diode detector leakage offset drop with frequency resulting in a greater power levels at the lower half of the spectrum. The five spiral resonances are detected correctly, therefore authenticating the correct operation of the transceiver.

After achieving satisfactory results from the wired testing of the "00000" tag, tags with different encoded data –11111 and –10101– are also examined using the same transceiver setup as shown in Fig. 6.22. The calibrated 8-bit digital spectral signatures of the three tags (00000, 10101 and 00000) are plotted in Fig. 6.24.

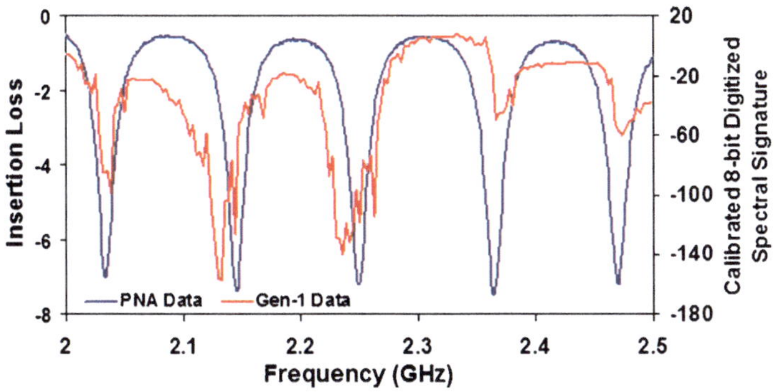

Fig. 6.23 Measured spectral signature of multiresonator using Agilent's PNA and Gen-1 transceiver

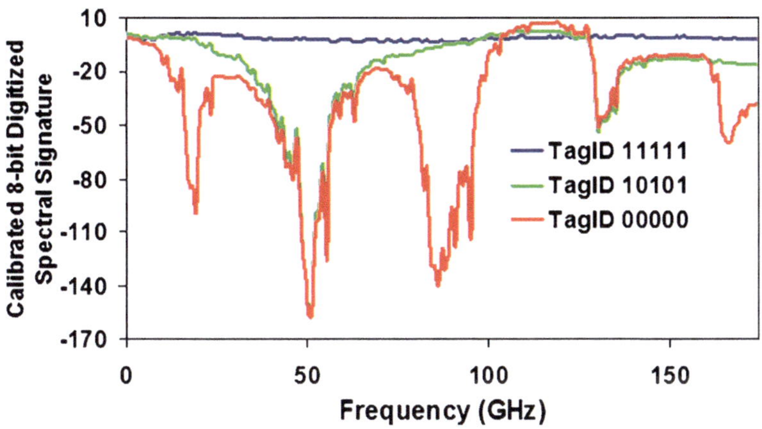

Fig. 6.24 Different spectral signatures of multiresonator detected using Gen-1 transceiver

As can be seen, tag 11111 gives no nulls as expected, whereas 10101 yields two nulls and 00000 yields five nulls at the respective resonant frequencies. The distinction in data sets from the three different tags proves the viability of the chipless stag concept.

The following section will present the results of the Gen-2 transceiver.

6.5.2 *Gen-2 Transceiver*

The Gen-2 transceiver is tested by connecting the tag multiresonator to the Tx and Rx ports as presented in Fig. 6.10. The Gen-2 transceiver is designed to detect the amplitude and phase of the spectral signature using a gain/phase detector AD8302 at the receiver end.

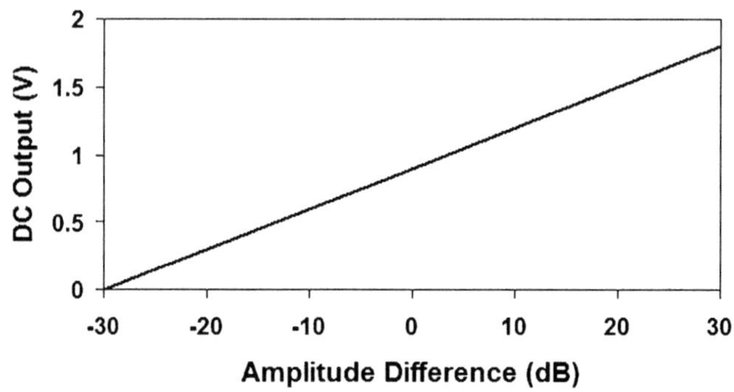

Fig. 6.25 Measured AD8302 DC output based on amplitude difference

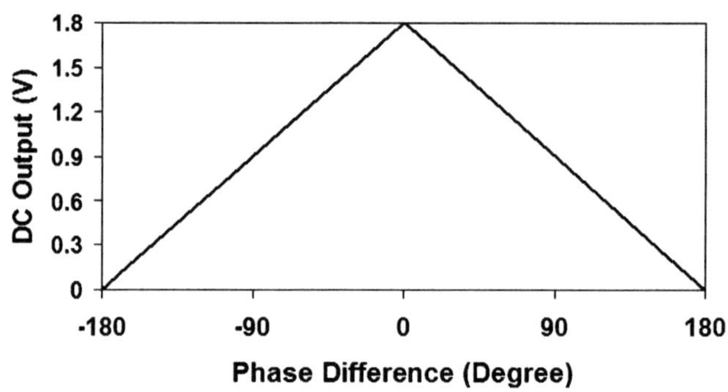

Fig. 6.26 Measured AD8302 DC output based on phase difference

The 5-bit multiresonator is interrogated by varying the frequency of the CW signal from 2 to 2.5 GHz. The gain/phase detector requires two RF inputs in order to detect the phase and amplitude difference between the two RF signals. One RF input signal is the received signal passing through the multiresonator, while the second RF input is directly from the transmitter VCO and acts as a reference signal. The amplitude and phase difference of the two RF signals are displayed as different DC voltage outputs. The DC voltage relationships with the amplitude and phase differences between the two RF signals are shown in Figs. 6.25 and 6.26. It is clear that the AD8302 has a 60 dB dynamic range for amplitude difference detection. The phase difference detection is performed in such a way that a difference of ±180° gives a DC output of 0 V, while in phase signals give a voltage output of 1.8 V by the AD8302.

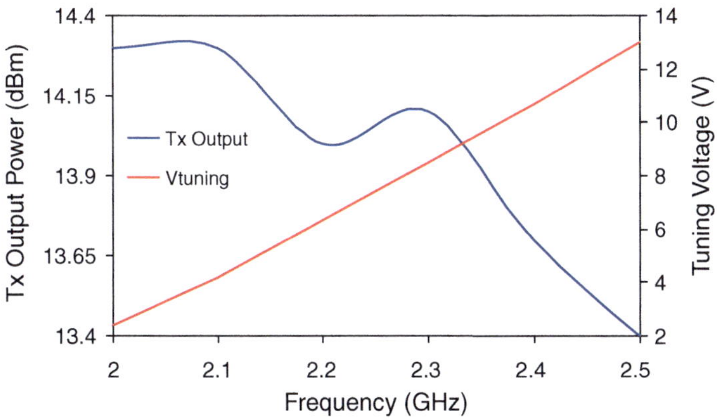

Fig. 6.27 Measured Gen-2 transmitter output power and VCO-tuning voltage

Following the successful design and testing of the Gen-1 reader, the testing of the Gen-2 transceiver is performed. The parameters that are characterized to analyse the Gen-2 transceiver performance are:

- Transmitter output power.
- Transmitting signal higher order harmonic distortion.
- Transmitter/receiver isolation.
- Phase error.
- Receiver sensitivity.

The measured Gen-2 transmitter output power is shown in Fig. 6.27. The RF output power is measured using Agilent's E8360 Spectrum Analyzer. The output power of the Gen-2 receiver is 3.3–3.5 dB less than the output power of the Gen-1 transmitter due to the power divider implemented after the VCO in the Gen-2 transmitter to provide a coherent reference signal for the Gen-2 receiver amplitude/phase detector. The tuning voltage sent to the Gen-2 transmitter VCO is presented in Fig. 6.27 and shows that the frequency control is very linear. Figure 6.27 shows that the transmitter output varies from 14.3 dBm at 2.1 GHz to 13.4 dBm at 2.5 GHz.

Figures 6.15–6.17 show the measured spectrum of the Gen-1 transmitter output from 0 to 10 GHz when the transmitter output signal frequency is set to 2, 2.25 and 2.5 GHz, respectively. The Gen-2 transmitter is designed the same as the Gen-1 transmitter except for the power divider circuit after the Gen-2 VCO. The power divider is a purely passive circuit, which does not produce any intermodulation distortion products. Hence, the higher order distortion measurements are the same with the Gen-2 transmitter as they are with the Gen-1 transmitter.

The measured isolation between the transmitter and receiver is shown in Fig. 6.28. The Gen-2 Tx/Rx isolation is similar to the Gen-1 Tx/Rx isolation due to the similar transceiver architecture. The experimental setup for measuring the Tx/Rx isolation was discussed in Sect. 6.5.1. From Fig. 6.28, it is clear that the isolation is above 71 dB from 2 to 2.5 GHz.

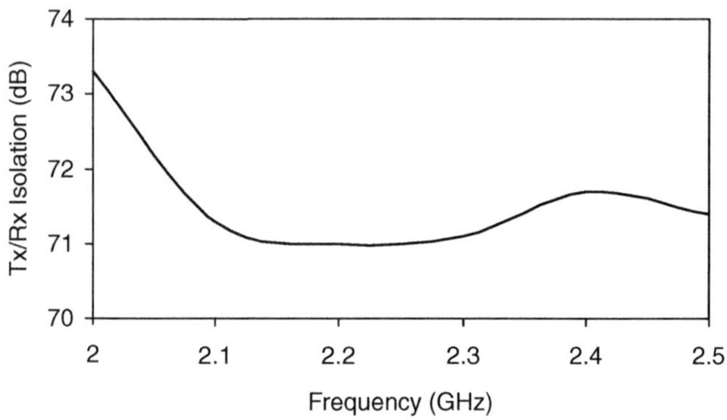

Fig. 6.28 Measured Gen-2 transmitter/receiver isolation

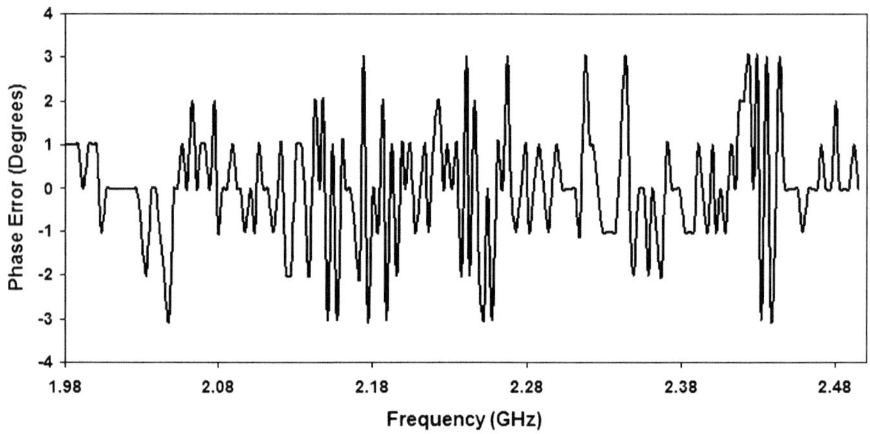

Fig. 6.29 Measured Gen-2 transceiver phase error

The Gen-2 receiver detects the amplitude and phase variations in the received tag signal by comparing it to a reference signal provided by the transmitter VCO. It is important to identify that the phase spectral signature encoding by the tag is more resilient to noise, but designing a receiver with phase data detection is a greater challenge. The coherent transceiver architecture of the Gen-2 is characterized by the phase error, which is present due to the phase noise from the VCO and nonlinearity of the power amplifiers in the receiver circuit. The measured phase error of the Gen-2 receiver is in the boundaries of ±3° and is shown in Fig. 6.29. The error is in reasonable boundaries since the tag decodes phase data variations in the ranges above 20°. Hence, the phase error induced does not influence the correct detection of the phase spectral signature.

The receiver sensitivity was tested using the experimental setup used for the Gen-1 as shown in Fig. 6.21. Agilent's E8257D signal generator was used to determine the minimal signal power levels that can be detected between 2 and 2.5 GHz.

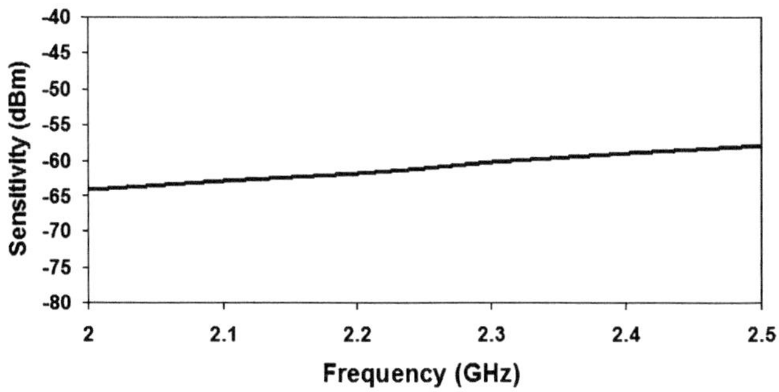

Fig. 6.30 Measured Gen-2 receiver sensitivity

Table 6.7 Gen-2 transceiver pre-design and tested specifications

Electrical specifications		
Specifications	Pre-design	Tested
Frequency of operation (GHz)	2–2.5	1.98–2.5
Transmitting power (dBm)	15	Above 13.4
Interrogation signal type	Frequency sweep	Frequency sweep
Tx/Rx isolation (dB)	60	Above 71
Receiver sensitivity (dBm)	−35	Below −57.9
Max. power consumption (Watts)	2	1.6
Commercial		
Cost	Less than AU $200 (a guide only)	$135

The receiver sensitivity determines the minimal received signal input power that can be successfully received by the Gen-2 transceiver. Figure 6.30 shows the measured Gen-2 receiver sensitivity. From Fig. 6.30, it is clear that the receiver sensitivity is below −57.9 dBm. The Gen-2 receiver sensitivity is increased in comparison to the Gen-1 since the AD8302 gain/phase detector circuit has greater sensitivity than the diode detector circuit.

The Gen-2 transceiver characteristics are summarized in Table 6.7 along with the minimum specifications set in the design section. From Table 6.7, it is clear that the Gen-2 transceiver has met all of the desired requirements accept for the transmitter output power. Since the receiver sensitivity is greatly increased than the specified −35 dBm, this will not be a problem. Thus, the design and testing of the Gen-2 transceiver is finalized.

The experimental setup of the transceiver integrated with the control section is shown in Fig. 6.31.

The Agilent PNA E8361A network analyser is used to measure the spectral signature amplitude (insertion loss) and phase (transmission phase) of the 5-bit multiresonator. Figures 6.32 and 6.33 show the measured spectral signature 000000 in amplitude and phase obtained from the PNA against the detected calibrated spectral

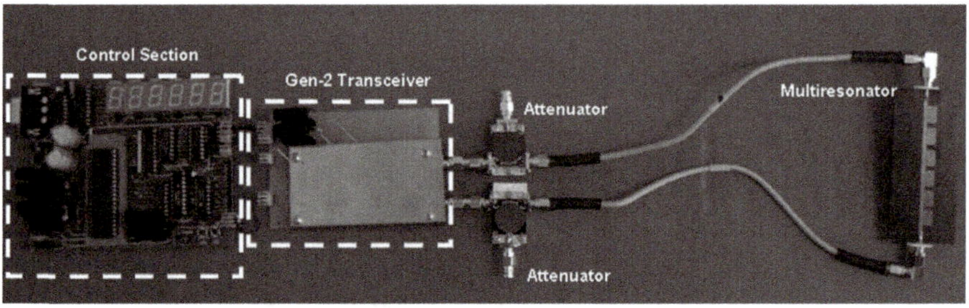

Fig. 6.31 Photograph of Gen-2 testing experimental setup for wired 6-bit tag detection

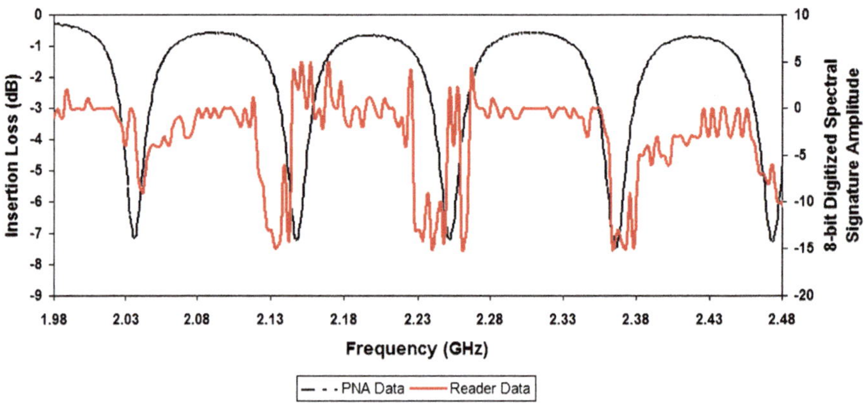

Fig. 6.32 Measured amplitude spectral signature using Agilent's PNA and Gen-2 transceiver

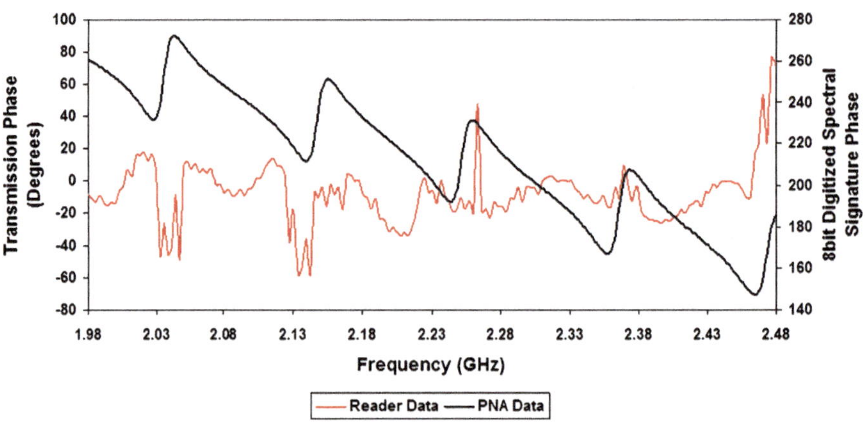

Fig. 6.33 Measured phase spectral signature using Agilent's PNA and Gen-2 transceiver

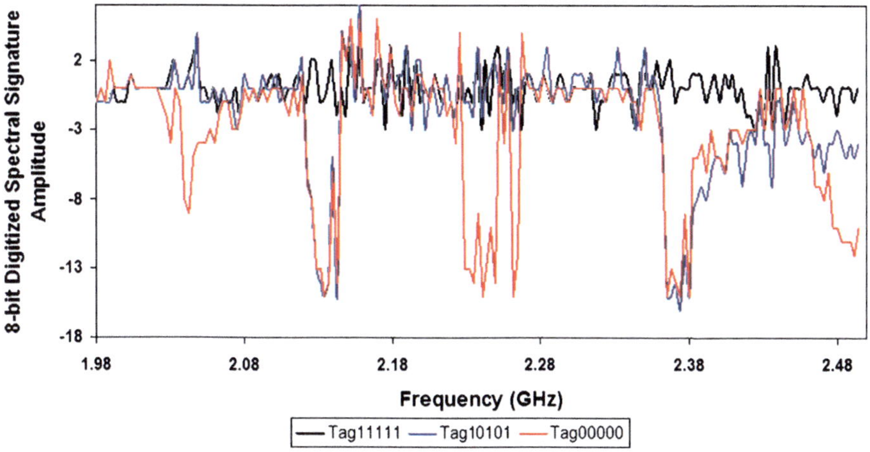

Fig. 6.34 Different measured amplitude spectral signatures using Gen-2 transceiver

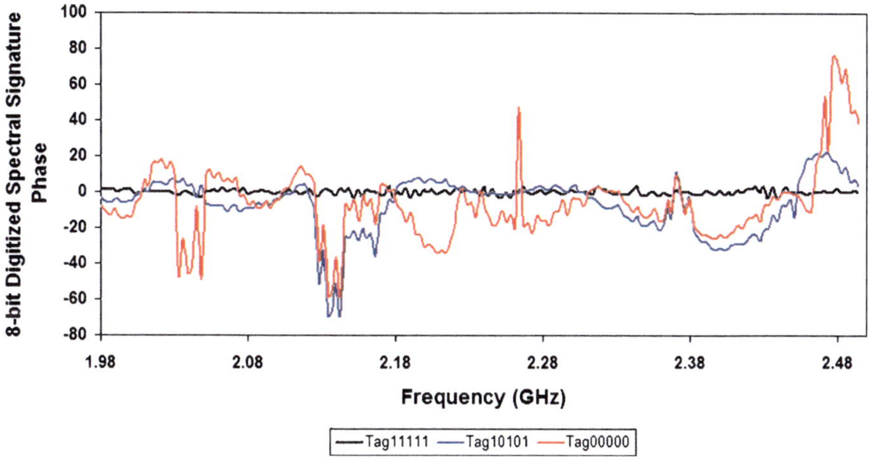

Fig. 6.35 Different measured phase spectral signatures using Gen-2 transceiver

signature using the Gen-2 transceiver, respectively. As with the Gen-1 transceiver testing, a tag with no resonances (11111) was used as a reference tag for calibration. Figures 6.34 and 6.35 show the detection of different amplitude and phase spectral signatures using the Gen-2 transceiver. As can be seen, tag 11111 gives no nulls or phase jumps as expected, whereas 10101 yields two nulls and phase jumps and 00000 yields five nulls and phase jumps at the respective resonant frequencies. The distinction in data sets from the three different tags proves the viability of using phase data detection to verify the amplitude data since the phase data is more resistive to interference and noise.

The following section will present the results of the UWB transceiver.

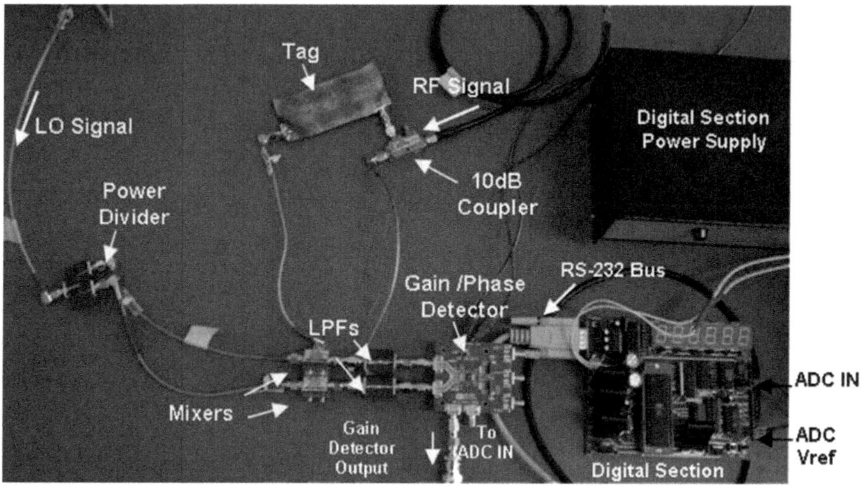

Fig. 6.36 Photograph of UWB transceiver experimental setup

6.5.3 UWB Transceiver

The UWB transceiver interrogates the tag by sweeping the RF signal from 5 to 9 GHz in order to detect the tag's 17 resonant nulls. The upper frequency was limited by the performance of the 10 dB coupler, which worked up to 9 GHz, and the mixers, which worked up to 10 GHz. The YIG oscillator generated the interrogation signal from 5 to 6 GHz, while the E8257D signal generator was used to generate the RF interrogation signal above 6 GHz. The tag multiresonator was wired to the transmitter and receiver as shown in Fig. 6.36. After the RF signal passed through the tag, it was encoded with the tags unique spectral signature ID and was then mixed to a lower IF frequency using ZX05-14+ mixer circuits at the receiver end. The LO signal was generated using RVC6000 VCO in the range of 4–8 GHz. The spectrum of the received tag signal after down-conversion at 1 GHz is shown in Fig. 6.37. The IF signal is then fed into the AD8302 gain detector for demodulation since it operates from 0.3 to 2.7 GHz as in the low frequency Gen-2 RFID transceiver.

The IF signal with the tags ID was compared against an IF reference signal and the magnitude and phase differences are given by the AD8302 as an analogue voltage output, which is sent to the digital/control board where it is digitized and further processed. A 17-bit multiresonator was used to test the UWB transceiver. The digitized 8-bit normalized amplitude and phase spectral signature of the 17-bit multiresonator is shown in Fig. 6.38. The 17-bit multiresonator data is normalized in respect to a tag with no resonances.

From Fig. 6.38, it is clear that the UWB transceiver operates correctly and detects 17 bits of data in both amplitude and phase of the spectral signature. The significance of the successful operation of the UWB transceiver is that the transceiver architecture which utilizes downconverting mixer circuits as an CW coherent radar

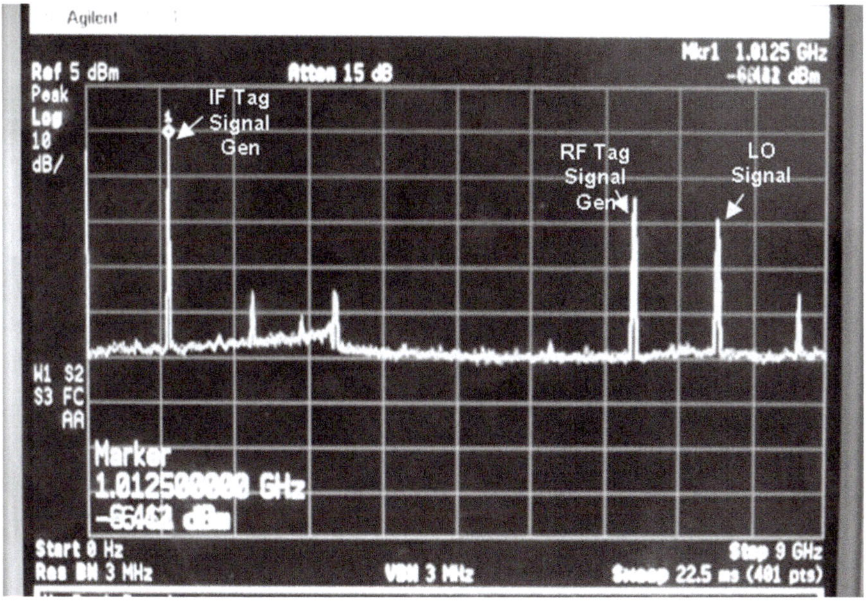

Fig. 6.37 Spectrum of received tag signal after down-conversion

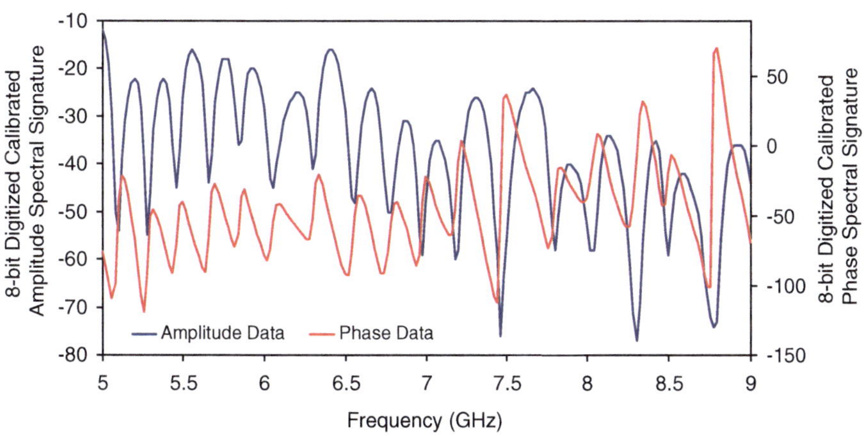

Fig. 6.38 Digitized amplitude and phase 17-bit spectral signature obtained using UWB transceiver

can be used within the entire UWB spectrum. Hence, a proof-of-concept transceiver circuit for the UWB chipless tag reader is successfully designed and tested.

The measured sensitivity of the UWB receiver is shown in Fig. 6.39. From Fig. 6.39, it is clear that the sensitivity of the receiver is increased in comparison to Gen-1 and Gen-2 circuits. The increase in sensitivity is due to the higher gain of the amplifiers in the receiver and the increased Tx/Rx isolation, which is shown in

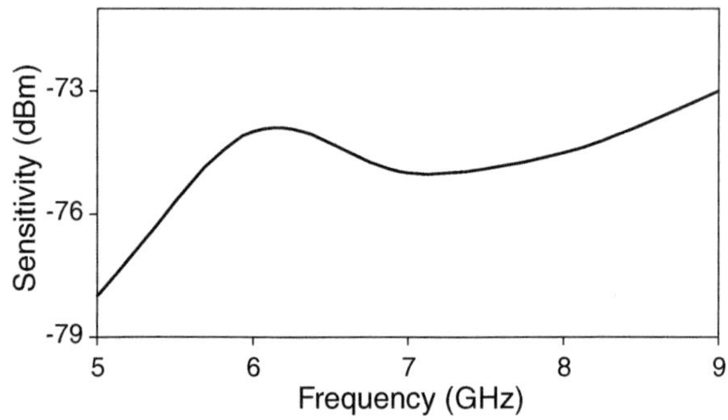

Fig. 6.39 Measured sensitivity of UWB receiver

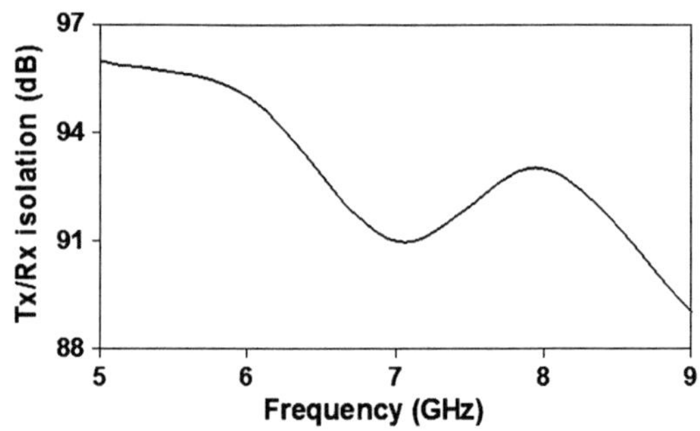

Fig. 6.40 Measured transmitter/receiver isolation of UWB receiver

Fig. 6.40. However, the sensitivity is limited to −78 dBm due to the large bandwidth of the receiver and the limitation of the AD8302 gain/phase detector which can only handle signals above −60 dBm. The measured sensitivity and Tx/Rx isolation of the receiver circuit satisfies the requirements set in the pre-design stage.

The YIG oscillator, which was used to generate the interrogation signal up to 6 GHz (and was replaced with an Agilent signal generator for signal generation above 6 GHz), has an upper frequency limit of 6 GHz and a constant output power of 16 dBm. It also dissipates a significant amount of power (~8 W) and heat and was mounted on a large heat sink as shown in Fig. 6.41. For future designs, a YIG oscillator operating from 5 to 10.7 GHz can be used along with mixers that operate above 9 GHz. Although these two components did not satisfy the required bandwidth, the concept of the UWB transceiver circuit is proven and can be utilized for amplitude and phase spectral signature tag detection. The testing and characterization of the UWB transceiver with the pre-design requirements are summarized in Table 6.8. The successful design, testing and characterization of the UWB RFID transceiver for the UWB chipless tag RFID reader finalize the results section.

Fig. 6.41 UWB Teledyne YIG oscillator mounted on heat sink

Table 6.8 UWB transceiver pre-design and tested specifications

Electrical specifications		
Specifications	Pre-design	Tested
Frequency of operation (GHz)	3.1–10.7	5–9 (due to component limitation)
Transmitting power (dBm)	15	16
Interrogation signal type	Frequency sweep	Frequency sweep
Tx/Rx isolation (dB)	60	Above 88
Receiver sensitivity (dBm)	−35	−78 to −73
Max. power consumption (Watts)	10	8.6
Commercial		
Cost	Less than AU $2,000 (a guide only)	$4,500

6.6 Conclusion

In this chapter, the differences between conventional and chipless RFID readers, transceiver specifications, the design and testing of three transceiver topologies for the chipless tag RFID readers are presented. The three transceiver topologies that were presented, designed and tested were: Gen-1, Gen-2 and UWB transceivers.

The Gen-1 transceiver was designed as a proof-of-concept detection circuit for the 6-bit tag on Taconic PCB substrate. The transceiver was designed to operate between 2 and 2.5 GHz and detected the amplitude spectral signature data of the chipless tag using a diode detector circuit at the receiver. The transceiver performed a linear frequency sweep interrogation technique to detect the resonant nulls of the tag. Successful detection of the tag's spectral signature confirmed that the transceiver circuit was operational. Additional testing of the transmitter power output vs. frequency, receiver sensitivity and isolation between the transmitter and receiver were conducted and were within the pre-required limitations.

The Gen-2 transceiver was the second designed transceiver, which was designed to detect the amplitude and phase spectral signature of the 6-bit tag on Taconic PCB substrate. The transceiver was designed to operate between 2 and 2.5 GHz and detected the amplitude and phase spectral signature data of the chipless tag using a gain/phase detector AD8302. The Gen-2 transceiver architecture was based on coherent CW radar in order to neutralize any phase errors or frequency drifts coming from the transmitter's VCO. The Gen-2 transceiver VCO performed a linear frequency sweep interrogation technique to detect the spectral signature of the tag. Successful detection of the tag's amplitude and phase spectral signature confirmed that the transceiver circuit was operational. Additional testing of the transmitter power output vs. frequency, receiver sensitivity and isolation between the transmitter and receiver were conducted and were within the pre-required limitations.

The third and final transceiver to be designed was the UWB transceiver. Since the chipless RFID tags are designed to operate in the UWB region, a transceiver topology, which can perform frequency sweep interrogation within the UWB spectrum with amplitude and phase spectral signature detection, was designed. The UWB transceiver was designed to operate between 5 and 9 GHz due to limitations based on the components that were used (YIG oscillator and mixers). The UWB transceiver successfully detected the amplitude and phase spectral signature data of the chipless tag using a gain/phase detector AD8302. The use of down-conversion mixers enabled the use of the AD8302 detector, which only operates up to 2.7 GHz. Hence, a RVC6000 VCO was used as a local oscillator to downconvert the received tag signal. The transceiver architecture was based on coherent CW radar in order to neutralize any phase errors or frequency drifts coming from the transmitter's YIG oscillator and LO VCO. The transceiver oscillator performed a linear frequency sweep interrogation technique to detect the spectral signature of the tag. Successful detection of the tag's amplitude and phase spectral signature confirmed that the UWB transceiver circuit was operational.

The use of phase data detection by a spectral signature detecting chipless tag RFID reader was the first of its kind to be investigated, designed and tested. The use of phase decoding along with amplitude data decoding allows for comparison between the two for more robust and accurate readings.

The following chapters will concentrate on the system integration of the chipless tag and readers in order to complete the system performance evaluation.

Chapter 7
Chipless RFID Tag-Reader System

Abstract In this chapter, we present the integration of the chipless RFID tag and RFID reader presented in the preceding chapters into a system. The implementation constraints in terms of reading range, tag orientation and operating frequency are specified. Specific components of the RFID reader's digital control section along with the tag ID detection/decoding algorithm are presented. Two chipless RFID systems are tested. First, a proof-of-concept system operating between 2 and 2.5 GHz with a 6-bit tag is tested. Following is the intended UWB chipless RFID system operating between 5 and 10.7 GHz. These tests finalize the system level integration and testing of the chipless RFID system.

7.1 Introduction

The preceding two chapters presented the design and results of the two main components of the RFID system: the chipless tag and the RFID reader transceiver. The final chipless RFID tag is fully printable and designed on thin flexible laminate using CPW technology. The tag is detected by an RFID reader. Three different transceiver circuits have been presented in the previous chapter, which interrogated the chipless tag by a frequency swept CW interrogation signal.

The system level integration of the chipless RFID tag and reader is presented in this chapter. The main goal of the system design is integration of the system components into an operational unit. In this particular case, the RFID reader is installed to detect and successfully decode the spectral signature of the chipless tag within the reader's interrogation zone.

RFID is an application-driven technology. Hence, the application defines the operational constraints in which the chipless RFID system is to operate. The main application for this chipless RFID system is to tag the flexible Australian polymer

S. Preradovic and N.C. Karmakar, *Multiresonator-Based Chipless RFID: Barcode of the Future*, DOI 10.1007/978-1-4614-2095-8_7,

banknote on a conveyor belt application. The polymer banknote is still attached to the substrate, which is also the conveyor belt which moves at a constant speed. The fact that the conveyor belt is the polymer banknote substrate itself means that every banknote (and therefore tag) is oriented the same way and with the same dimension and condition (flat, straight and not bent), which does not create any orientation and alignment problems.

As mentioned earlier, the two main system components are the chipless RFID tag and the RFID reader. Three different chipless tags are used to test the RFID reader circuits. The 6-bit proof-of-concept tag operating from 2 to 2.5 GHz is used to test the successful operation of the Gen-1 and Gen-2 reader. The UWB 23-bit CPW tag is used to test the UWB RFID reader circuit. The Gen-1, Gen-2 and UWB RFID reader transceiver circuits are integrated with the digital control section to form the RFID readers.

Firstly, the RFID system is tested in an anechoic chamber, which is an interference free environment, followed by a second test in a laboratory, which is a noisy environment. The chipless tag is detected at first by using the Agilent's E8361A PNA as the reader circuit with two cross-polarized reader antennas connected at the two ports of the PNA. Full two-port calibration from 2 to 2.5 GHz was performed before the test. The tag's spectral signature is detected as the insertion loss between the two ports of the PNA.

The successful chipless RFID system operation is confirmed by the successful detection of the predefined response (spectral signature) of the chipless tag by the developed RFID reader. Besides the hardware design of the tags and the reader, a Windows-based application software is developed in order to completely automate the RFID system. The application software is developed using Visual Basic 6 in order to plot the received tag's spectral signature in both amplitude and phase sent by the reader. The software application controls the RFID reader by sending calibration, single read (polling) and continuous reading of chipless tags.

The chapter is organized as follows: Section 7.1 has presented the RFID system overview. Section 7.2 presents the application and the implementation constraints of the system to address the specific application for banknote tagging. Section 7.3 presents the description of hardware and software components to be used in the proposed system. Following are the field trials with setup and results obtained for three varieties of reader presented in Sect. 7.4. Finally, the conclusions are presented in Sect. 7.5. The following section presents the chipless RFID system application and implementation constraints.

7.2 Application and Implementation Constraints

In this section, the application for which the RFID system has been designed for and the minimum system performance specifications that are needed to be met are presented. The chipless tag is designed to be printed on the Australian polymer banknote. The RFID reader is designed to detect the tag at short ranges and to be

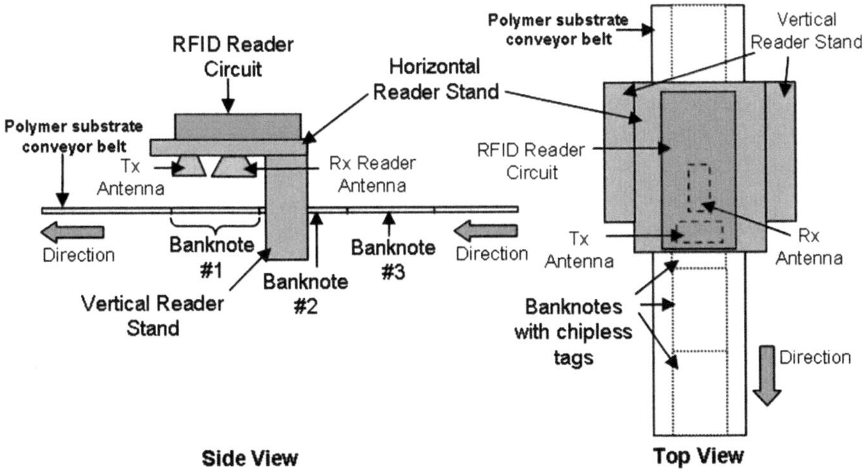

Fig. 7.1 Block diagram of conveyor belt application for chipless RFID system

mounted over a conveyor belt. The important system constraints and installation requirements are identified and they are:

- Tag alignment.
- Minimum required reading range.
- Tag reading rate.

In the preceding chapters, it has been clearly stated that the chipless RFID system utilizes two cross-polarized tag antennas and corresponding reader antennas in order to isolate the interrogation signal and the tag's response signal. This requirement put stringent alignment and polarization prerequisites for the tag and the reader antennas. Since the RFID installed on a conveyor belt system is intended for single tag readings at a time, alignment and polarization requirements must be met for the system to operate properly. It is important to remember that the banknotes are intended to be tagged while still part of the banknote substrate which is the conveyor belt itself. Hence, every tag is printed correctly and exactly the same way on a different banknote, with correct polarization, same separation between tags and moving at a constant speed. Figure 7.1 shows the block diagram of the conveyor belt application of the RFID system for banknote tagging.

The maximum reading range of the reader is limited by the reader's sensitivity and emitted EM power. However, for the successful detection of the tag printed on the banknote (substrate) at a great reading range is not required. Hence, a nominal 10 cm reading range is nominated as the minimum reading range.

The tag reading rate is a property of the RFID reader circuit. The reader needs to interrogate, detect and decode the tag's identity before the next tag enters the reader's interrogation zone. The highest read rate of the reader is determined by two factors:

- Minimum signal bandwidth.
- Digital circuit processing capabilities.

The interrogation of the tag is performed using a linear frequency sweept CW signal. The duration of the transmitted interrogation signal determines its bandwidth, i.e. a 1 μs CW signal has a bandwidth of 1 MHz. Hence, the desired spectral resolution determines length of the interrogation signal. For a 1 MHz resolution over a 5.7 GHz bandwidth, the interrogation time would be $5{,}700 \times 1\ \mu s = 5.7$ ms. This means that 175 tags per second could be read with a resolution of 1 MHz. This means that the reader would be operational for a conveyor belt moving at approximately 1 m/s (60 m/min). It is important to mention that a 1 MHz resolution could not be needed, which means that the interrogation time would decrease and hence the conveyor belt speed could increase if needed.

Another important limiting factor is the electronics that are used for the reader supporting the interrogation speed. Most reduced instruction set computer (RISC) microprocessors can operate with an instruction cycle of 100 ns, while digital signal processors (DSP) operate with instruction cycles of up to 1 ns.

Hence, we can conclude that the processing capabilities and speed of today's microprocessors/DSP's are sufficient to cater the interrogation of the banknotes on a conveyor belt.

The following section presents the component of the chipless RFID system.

7.3 Chipless RFID Tag-Reader System Components

The system specifications and conveyor belt application were presented in the preceding section. Two chipless RFID systems with components are presented in this chapter. The first system is a 6-bit proof-of-concept chipless RFID system, while the second RFID system is a UWB chipless RFID system.

The 6-bit proof-of-concept chipless RFID system operates from 2 to 2.5 GHz. The system components of the proof-of-concept chipless RFID system are shown in Fig. 7.2. The proof-of-concept chipless RFID system comprises a 6-bit chipless RFID tag, which is detected either with a Gen-1 RFID reader (amplitude only spectral

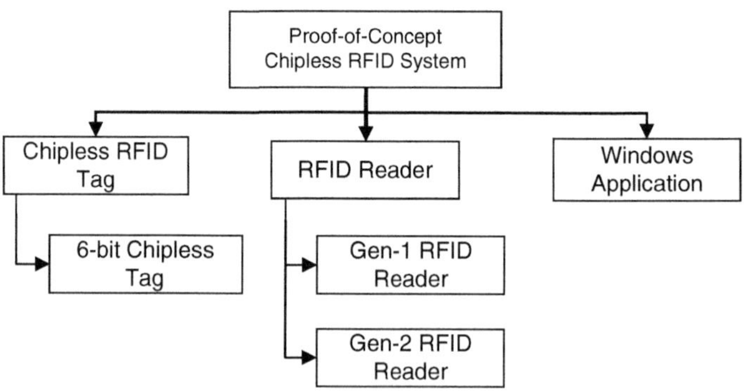

Fig. 7.2 Block diagram of 6-bit proof-of-concept chipless RFID system components

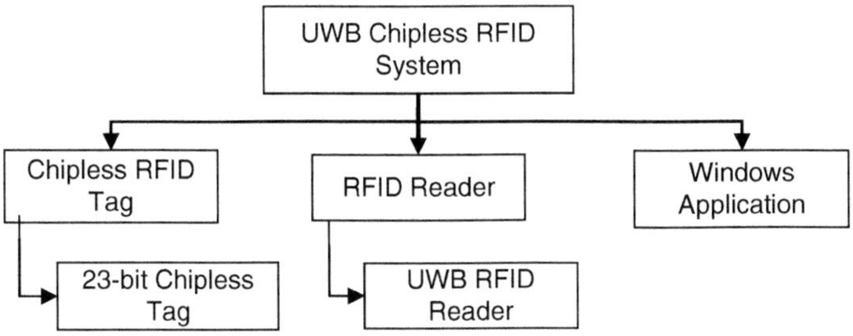

Fig. 7.3 Block diagram of UWB chipless RFID system components

signature detection) or a Gen-2 RFID reader (with amplitude and phase spectral signature detection), and a Windows-based software application, which controls the RFID reader via an RS-232 bus and displays the received tag data.

The UWB chipless RFID system operates from 5 to 10.7 GHz. The system components of the UWB chipless RFID system are shown in Fig. 7.3. The UWB chipless RFID system comprises a 23-bit chipless RFID tag, which is detected using an UWB RFID reader, and a Windows-based software application, which controls the RFID reader via RS-232 bus and displays the received tag data.

The following sections present the chipless tags, RFID readers and applications of the two chipless RFID systems.

7.3.1 Chipless RFID Tags

Two chipless tags are used in the reading process: 6-bit microstrip chipless tag on standard PCB laminate Taconic TLX-0 and 23-bit CPW chipless tag on thin and flexible laminate Taconic TF-290. The design and performance of the chipless tags has been presented comprehensively in Chap. 5 and are shown in Fig. 7.4.

7.3.2 RFID Reader Digital Control Section

The digital section is designed around an 8-bit Atmel AT89C52 microcontroller, which performs the major signal processing and data-decoding algorithms. The architecture of the digital section is presented in Fig. 7.5. A microcomputer system has been used. The whole concept is based on using a CPU (AT89C52) and the peripheral components with specific external memory address allocations. The microcontroller communicates with the peripheral components via an 8-bit digital data bus and an 8-bit address bus. The data bus is connected to the 8-bit input ports of DAC, ADC and display buffers. The address bus is sent to the address decoder, which then determines which peripheral unit to activate. Each peripheral unit has a chip select (CS) pin, which activates the specific IC. After activating the peripheral unit, data is put on the data bus and the write signal is clocked, hence loading the IC

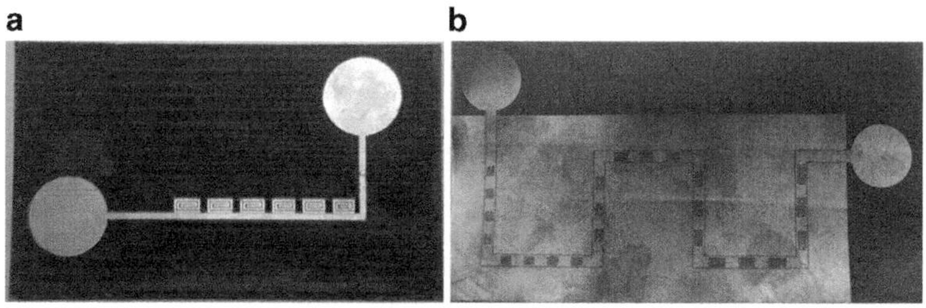

Fig. 7.4 Chipless RFID tags: (**a**) microstrip tag with 6 bits of data and (**b**) CPW tag with 23 bits of data

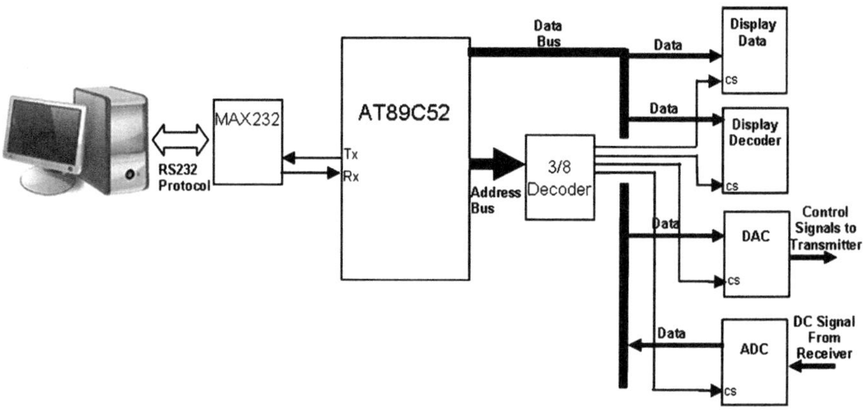

Fig. 7.5 Block diagram of chipless RFID reader digital control section

with data from the CPU through the data bus. The photograph of the digital section is shown in Fig. 7.6. The components used for the short range RFID reader are:

- AT89S52 8-bit CMOS microcontroller—Atmel.
- AD7533JN DAC—Analog Devices.
- AD817AN High Speed Amplifier—Analog Devices.
- AD7819 ADC—Analog Devices.
- MAX232 RS232 Driver-Maxim-Dallas.
- 6 Seven Segment Displays Common Anode.
- Buffers and DFF Latches—Texas Instruments.

The digital board was upgraded with a LCD display, keyboard and two 10-bit ADC instead of a single 8-bit ADC so that both amplitude and phase difference could be sampled simultaneously at a higher resolution. The upgraded digital section is used for the UWB RFID reader since it requires more frequency points and higher resolution due to its ultra-wide bandwidth operation. The photograph of the upgraded digital control section is shown in Fig. 7.7.

The following section presents the integrated RF transceiver and digital section for the Gen-1, Gen-2 and UWB RFID reader device.

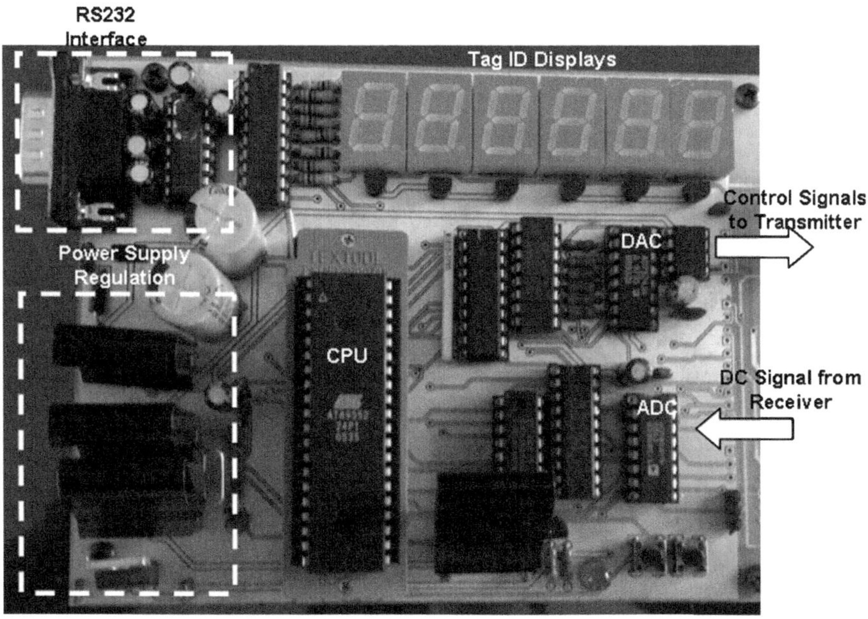

Fig. 7.6 Photograph of RFID reader digital section

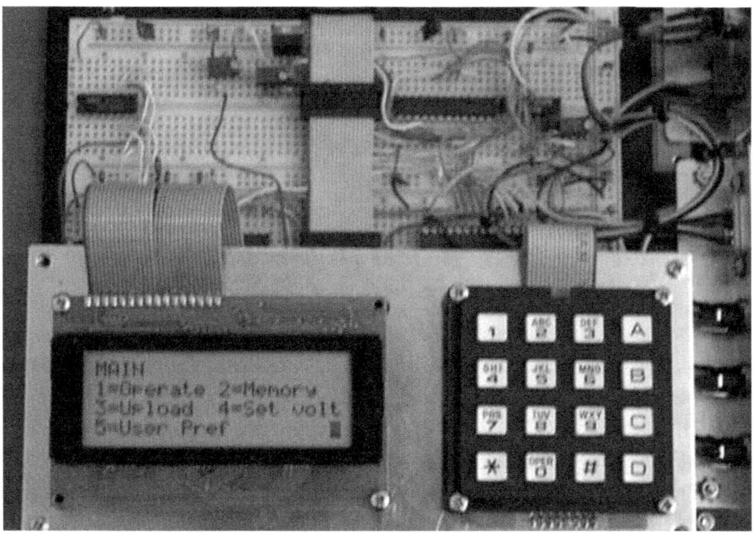

Fig. 7.7 Photograph of upgraded digital/control section for UWB RFID reader

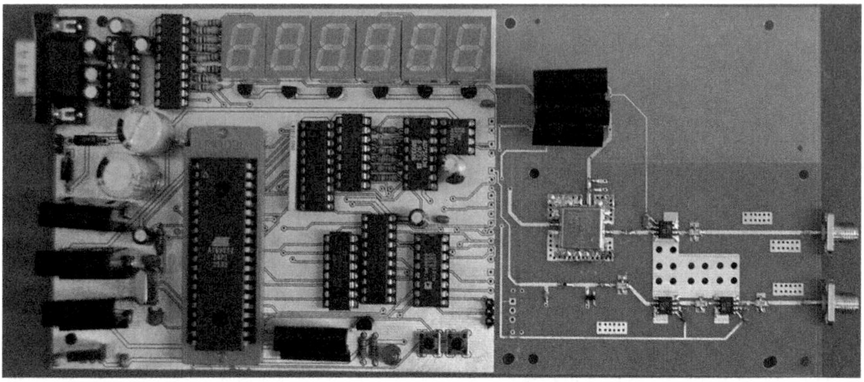

Fig. 7.8 Photograph of Gen-1 RFID reader with diode detector

Fig. 7.9 Photograph of Gen-2 RFID reader with gain/phase detector

7.3.3 Chipless Tag RFID Reader Devices

Three transceiver architectures for Gen-1, Gen-2 and UWB readers have been presented in Chap. 6. In this section, the integrated hardware of all three RFID readers (consisted of both digital and RF sections) is presented. The Gen-1 and Gen-2 RFID reader are shown in Figs. 7.8 and 7.9, respectively. These two readers are used for field trials of the 6-bit proof-of-concept 6-bit chipless RFID system. The Gen-1 and Gen-2 RFID reader are designed on low-cost FR4 ($\varepsilon r = 4.4$, $h = 0.5$ mm, $\tan\delta = 0.02$) laminate.

The photograph of the UWB RFID reader is shown in Fig. 7.10. This reader is used to perform field trials for the UWB chipless RFID system operating from 5 to 10.7 GHz.

The following section presents the interrogation/detection algorithm implemented in the three RFID readers. The algorithm is implemented using C code.

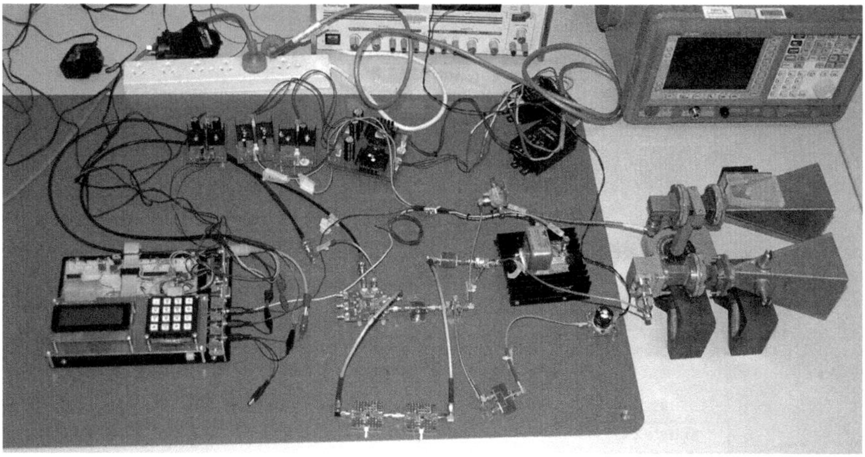

Fig. 7.10 Photograph of UWB RFID reader circuit

7.3.4 Chipless RFID Reader Tag Interrogation/Detection Algorithm

The Gen-1 and Gen-2 RFID readers interrogate the chipless tag by sweeping the frequency spectrum (2–2.5 GHz) in 150–180 samples (amplitude data in 180 points and phase data in 150 points). The UWB RFID reader sweeps the 5–10.7 GHz spectrum in 1,025 samples (5 MHz resolution). The interrogation/detection algorithm implemented in all three RFID readers is presented in Fig. 7.11. It was necessary to calibrate the RFID readers first by setting up the experiment and then determining the necessary amplitude and phase data threshold, which determined logics “1” and “0”. This was done by interrogating a tag with all logic zeros in its ID and recording the data and then replacing the tag with all logic ones in its ID. Hence, a clear difference between what is logic “0” and what is logic “1” is created and recorded in the reader. The calibration procedure enables the detection of a tag in the reader’s interrogation area by measuring the received signal strength.

Tag interrogation starts when the reader detects the received signal strength from a tag in its interrogation zone. After setting the interrogation frequency by loading the DAC with an 8-bit digital number corresponding to the necessary analog-tuning voltage for the VCO, the reader software reads data from the ADC. This data is the digitized values of the analog DC signals created by the gain/phase detector when comparing the received signal from the tag to the reference signal. The reader continues interrogating the tag until it reaches the final frequency sample. At this point, all the necessary data is collected and the tag ID decoding commences. The tag ID is decoded by using the threshold values established in the calibration routine. It is not necessary for the tag to be in the interrogation zone of the reader in order to perform this operation. After the tag ID has been decoded, it is sent to the display

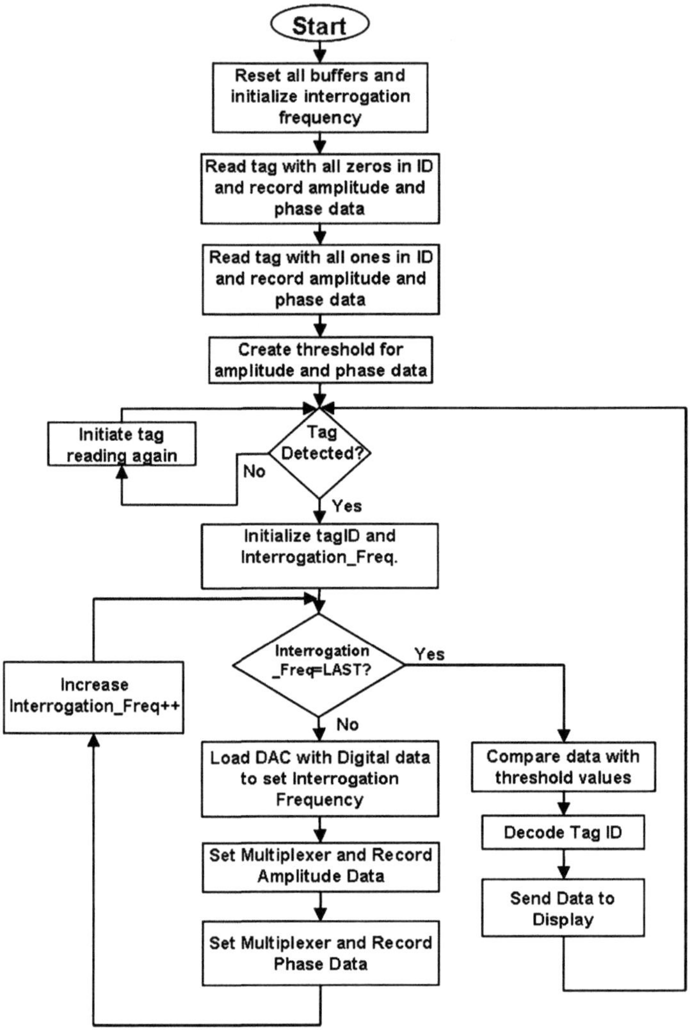

Fig. 7.11 Flow chart of the RFID reader ID decoding algorithm

and/or via RS-232 to the host computer application. The algorithm returns to Tag Interrogation mode and waits to detect the presence of the next tag.

This section finalizes the RFID reader hardware and software design. The following section presents the application software on a PC.

7.3.5 *Application Software for Chipless RFID System*

The application software is designed using Visual Basic 6 and operates on Windows XP operating system on a PC. The PC application software is used for automated

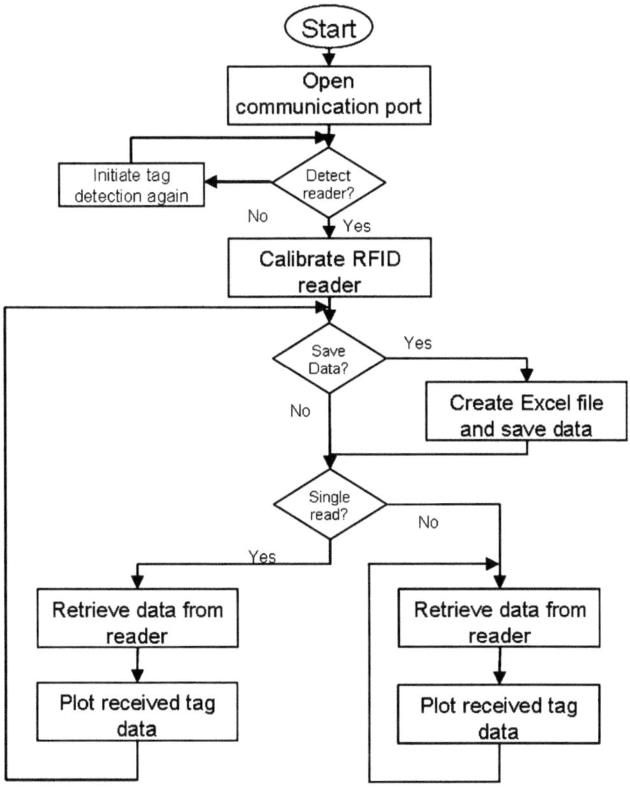

Fig. 7.12 Flow chart of the PC software application algorithm

data extraction from the reader. This means that there is no need for human interaction for the reader to interrogate the tag, send data to the PC application software using the RS-232 protocol where the tag ID and/or the tag data is presented. The application software algorithm is shown in Fig. 7.12.

The application software starts by opening the port on the PC (COM2) and detects the RFID reader. If no RFID reader I detected, a warning is sent to the user. Once the reader is detected at the COM port, the application software sends an instruction to the reader for calibration. The calibration of the reader is performed by reading a tag with no resonances (all ones), which eliminates the influences of the environment. Once the reader is calibrated, the application software does not require any more calibration from the reader. The system is then set in tag detection mode. At any time, the user can choose to save the tag data received by the reader in an Excel spreadsheet. If this choice is made, a new Excel file opens for a tag detection made by the reader and saves the data at a user-defined address on the PC.

The reader operates in single as well as continuous tag interrogation. This is predetermined by the user by making this choice on the application software menu. A single read option will require a tag reading/detection to be defined by the user.

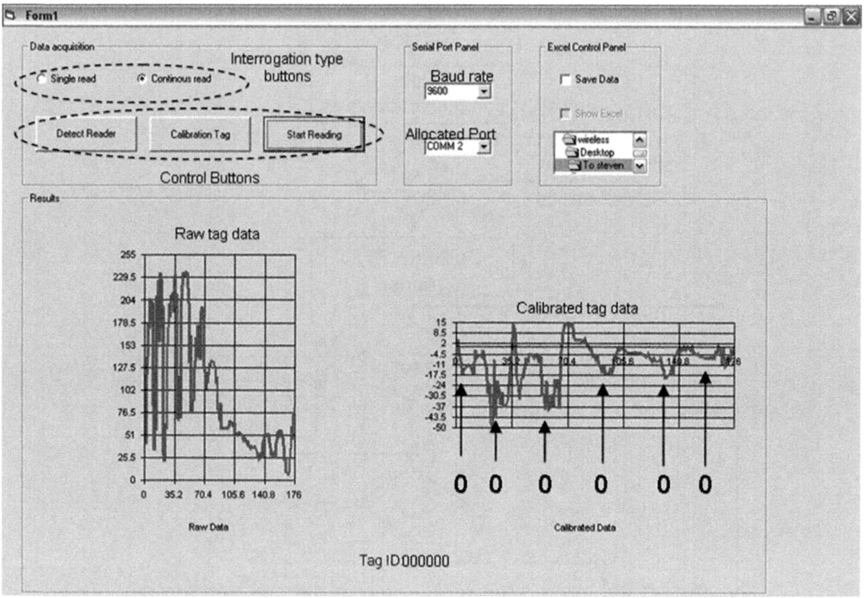

Fig. 7.13 Screen capture of PC software application

The continuous tag detection will insure that the reader is constantly reading and detecting the tags. This type of interrogation is intended for the conveyor belt application where the tags are constantly moving under the reader antennas. The single tag reading is used for system testing and field trials. Figure 7.13 shows the screen capture of the application software on the PC. The following section will present the chipless RFID system field trials.

7.4 Field Trials

In this section, the field trials of the chipless RFID tags and readers developed in this project are presented. Firstly, the 6-bit proof-of-concept chipless RFID tag and Gen-1 and Gen-2 RFID readers are tested inside an anechoic chamber (interference free environment) and outside the anechoic chamber (interference environment). Finally, the UWB RFID chipless tag and UWB RFID reader are tested inside and outside the anechoic chamber. Both systems are tested for reading ranges with amplitude and phase spectral signature detection. The amplitude and phase spectral signature detection are compared in order to establish robustness of the system. Figure 7.14 shows the block diagram of the field trials that are conducted for the 6-bit and 23-bit chipless RFID system.

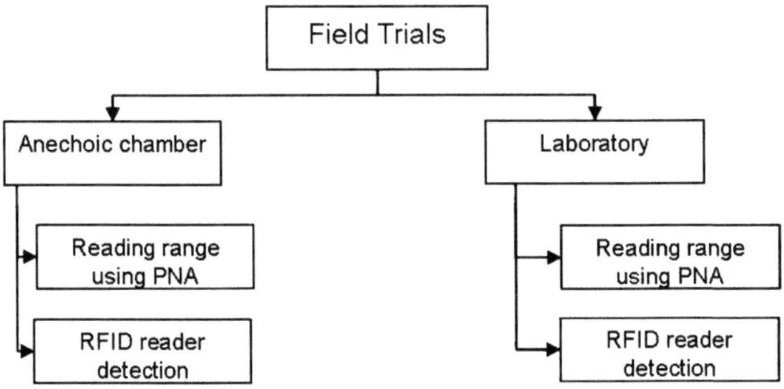

Fig. 7.14 Flow chart of the field trials conducted for chipless RFID system

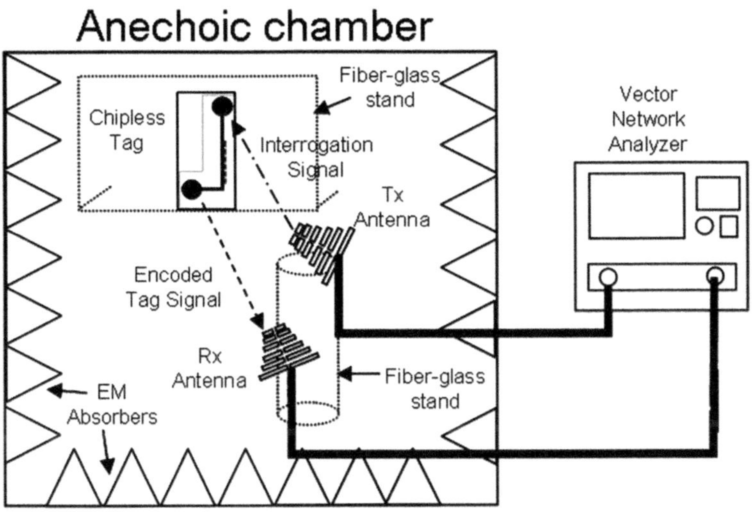

Fig. 7.15 Chipless RFID system block diagram

7.4.1 Six-Bit Proof-of-Concept RFID Chipless Tag: Reader System Field Trials

The experimental setup comprises the LPDA built on Taconic TLX-0, the vector network analyser E8361A as the reader electronics and the chipless tag. The block diagram and photograph of the experimental setup are shown in Figs. 7.15 and 7.16, respectively.

The experiment was conducted in an anechoic chamber in order to validate the successful encoding of the tag and its detection at the reader end using the network analyser. The chipless tag along with the reader antennas have been mounted on fiber-glass stands and placed into the anechoic chamber. Figure 7.16 shows that the

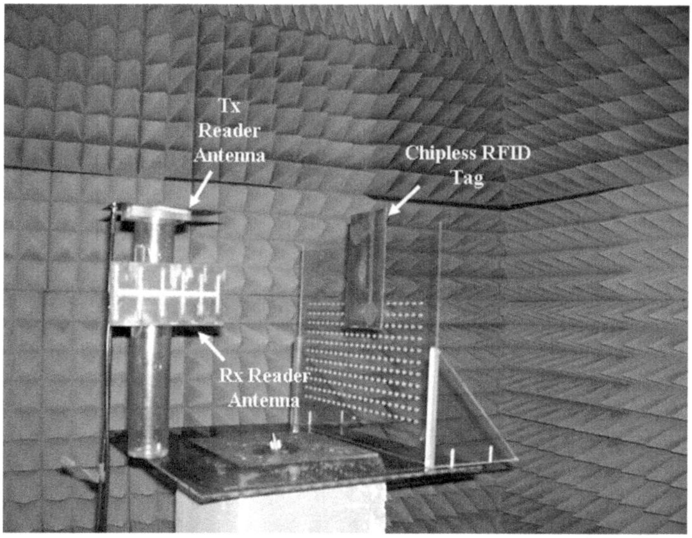

Fig. 7.16 Chipless RFID system experimental setup

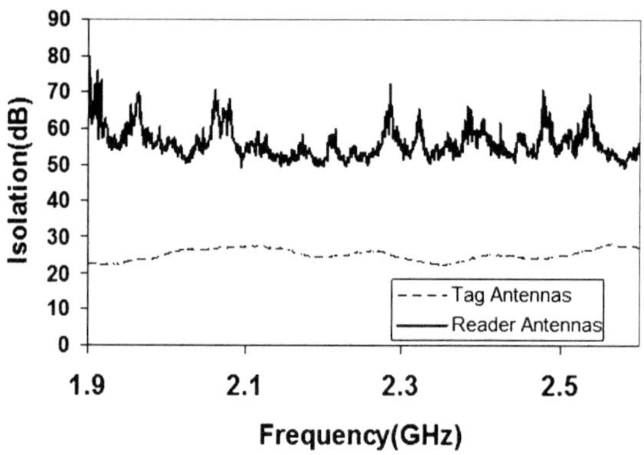

Fig. 7.17 Measured isolation between cross-polarized reader antennas

reader antennas are cross-polarized in order to minimize cross-talk between the two antennas, thus improving isolation of the tag signal. Figure 7.17 shows the measured isolation of the omni-directional UWB monopole tag antennas and the directive reader LPDAs. From Fig. 7.17, it is clear that the LPDAs provide greater isolation than the omni-directional monopoles between 2 and 2.5 GHz.

The LPDA is a preferred candidate for the reader antenna to the UWB monopoles due to better isolation between the two cross-polarized antennas (Chap. 4), more directive radiation pattern which creates a stronger line-of-sight (LOS) component, higher gain and its low cross-polar components.

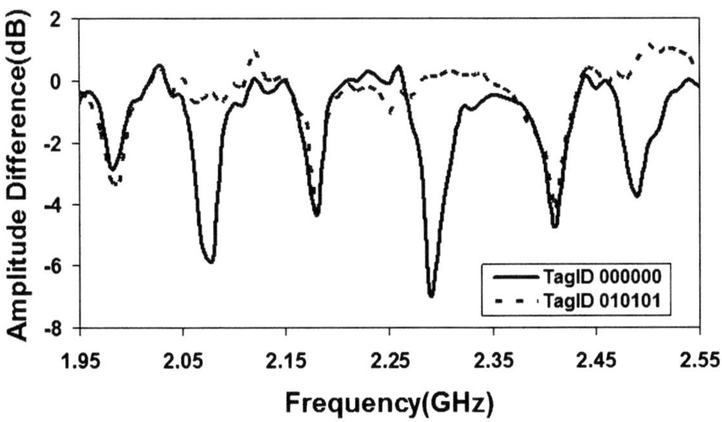

Fig. 7.18 Amplitude variations of the received tag signal at the reader end for chipless tags at 5 cm

Table 7.1 Amplitude and phase differences vs. reading distance of different bits of Tag000000 and Tag111111 measured in an anechoic chamber (ΔA—amplitude difference in dB and $\Delta\theta$—phase difference in degrees)

	2^5		2^4		2^3		2^2		2^1		2^0	
Distance (cm)	ΔA	$\Delta\theta$	ΔA	$\Delta\theta$	ΔA	$\Delta\theta$	ΔA	$\Delta\theta$	ΔA	$\Delta\theta$	ΔA	$\Delta\theta$
5	−3	14	−6	16	−4	17	−7	17	−4	17	−4	18
10	−1	15	−2	40	−3	30	−2	25	−1	15	−2	42
15	0	14	−1	18	−1	28	−1	16	−1	25	−1	15
20	0	12	−1	14	−2	17	−3	16	−1	26	−1	26
25	0	12	−4	16	−4	24	−3	22	−1	14	−3	23
30	0	10	0	18	−1	15	0	15	−1	21	−2	30
40	0	7	0	9	0	6	−1	12	0	12	−1	10

We encoded the tag with ID 000000 (reference tag) and placed it from 5 to 40 cm (in steps of 5 cm) away from the LPDA reader antennas as shown in Fig. 7.16. The same experiment was conducted using the UWB monopoles as reader antennas. However, a reading range of only a couple of centimeters was achieved. The measured results of the received data are presented in both magnitude and phase with the variation of the tag-to-reader distance. Agilent's E8361A network analyser was used in the measurement. The analyser was calibrated with the output power of the ports being −28 dBm.

The amplitude information in the spectral signature was measured and assessed in such a way that the tag with the 111111 ID is used as a reference. The amplitude difference of the received power due to the tag frequency signatures is presented in Fig. 7.18 and Table 7.1. Figure 7.19 shows the amplitude difference in respect to 111111 ID of the received signal after interrogating the same tag rotated by 180° hence creating a polarization mismatch. The amplitude information is considered completely unreadable and therefore useless. These results represent the robustness of the dual polarized reading system and the importance of alignment of the tag's antennas and the corresponding reader's antennas.

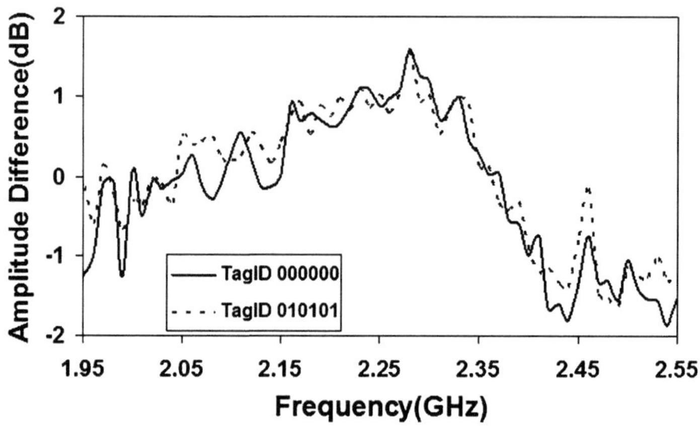

Fig. 7.19 Received signals by reader for chipless tags at 5 cm rotated by 180° (polarization mismatch between tag and reader antennas)

Table 7.1 shows that the tag readability drops with the reading distance. The tag was successfully read with LPDA at distances up to 10 cm, but reading ranges over 10 cm created errors mostly for 2^5 most significant bit which has the resonant null at ~2 GHz. These errors can be attributed to the high cross-polar level of the tag's monopole and impedance mismatch of both tag and reader antennas at that frequency. An optimization of the antenna design should rectify this problem.

The experiment was performed within the near-field and near far-field of the reader antennas (far-field is 38.4 cm at 2.25 GHz using $2d^2/\lambda$, where d is the antenna largest dimension in this case 16 cm and λ is wavelength). The system is hence designed to operate in both near-field and near far-field conditions.

The phase information in the spectral signature was measured and assessed in such a way that the tag with the 111111 ID is used as a reference. The phase variation of the received signal due to the tag frequency signatures is presented in Fig. 7.20 and Table 7.1. Figure 7.21 shows the phase information when the tag is rotated by 180° creating complete polarization mismatch with the reader antennas. The distinct phase difference between the two states is completely useless. Again, this result represents the robustness of the system, immunity to interference and the importance of alignment of the antennas of the tag and the reader.

From Table 7.2, we can see that the tag is accurately read by the reader (even at 40 cm which is the far-field), thus we can conclude that the phase information is more resilient to noise and can be read from a greater distance when compared to the amplitude information of the frequency signature. This represents a novelty in data encoding and data extraction in regards to other chipless spectral signature-based RFID systems. Previously reported spectral signature-based chipless RFID systems are based on examining and encoding the magnitude of the signal [18, 19].

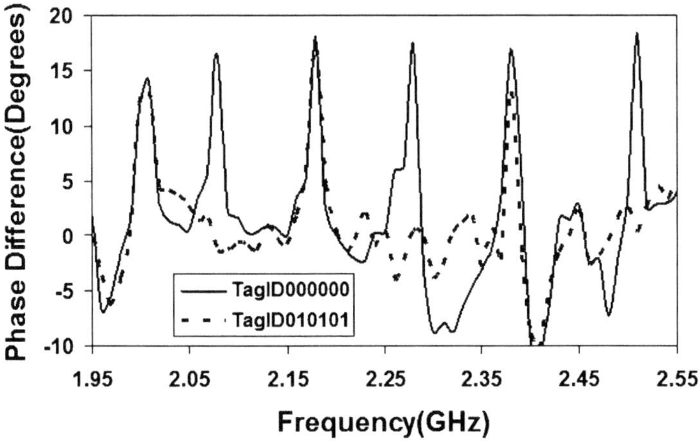

Fig. 7.20 Phase variations of the received tag signal at the reader end for chipless tags at 5 cm

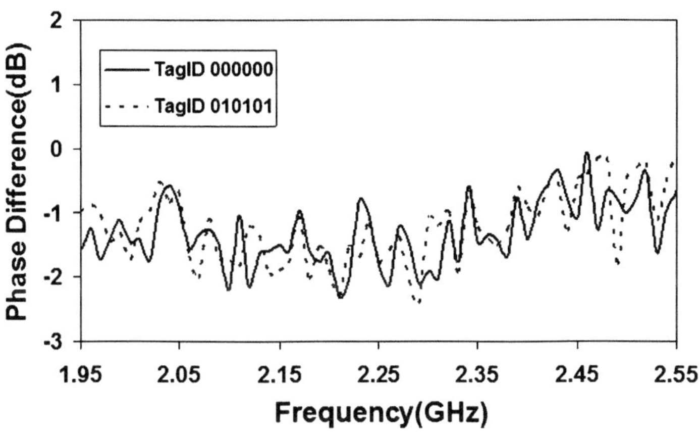

Fig. 7.21 Phase variations of the received tag signal at the reader end for chipless tags at 5 cm rotated by 180° (wrong alignment)

Table 7.2 Amplitude and phase differences vs. reading distance of different bits of Tag000000 and Tag111111 measured outside the anechoic chamber (ΔA—amplitude difference in dB and $\Delta\theta$—phase difference in degrees)

	2^5		2^4		2^3		2^2		2^1		2^0	
Distance (cm)	ΔA	$\Delta\theta$	ΔA	$\Delta\theta$	ΔA	$\Delta\theta$	ΔA	$\Delta\theta$	ΔA	$\Delta\theta$	ΔA	$\Delta\theta$
5	−2	17	−7	26	−4	30	−7	35	−5	28	−4	16
10	−2	17	−4	31	−2	22	−8	29	−4	39	−9	26
15	−1	25	−5	29	−3	18	−2	23	−2	36	−1	30
20	0	12	−2	19	−4	30	−3	20	0	11	−1	13
25	0	10	−4	16	−4	20	−3	25	−1	18	−3	26

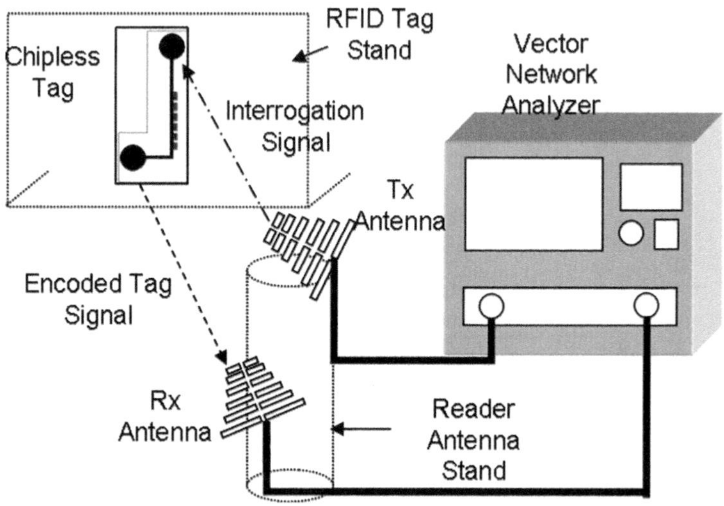

Fig. 7.22 Block diagram of the experimental setup using directive LPDAs as reader antennas

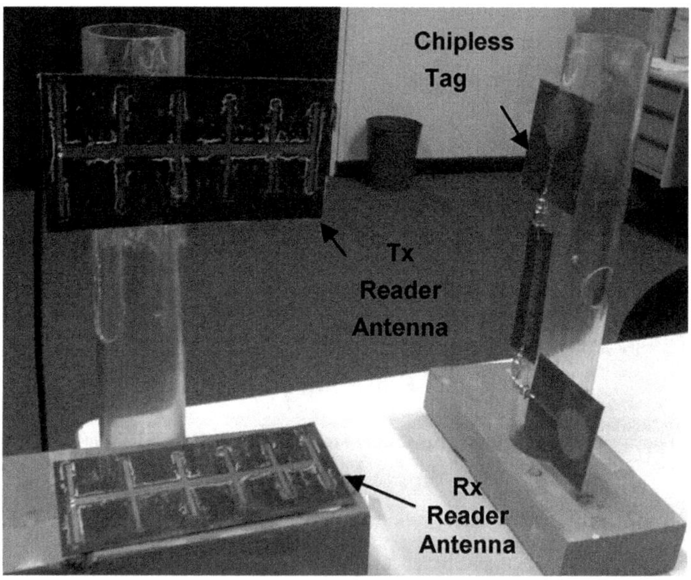

Fig. 7.23 Photograph of the experimental setup in the laboratory

Following the successful testing of the chipless RFID system inside the anechoic chamber, the chipless system was setup outside the chamber. The experimental setup comprises two cross-polarized LPDAs fabricated on Taconic TLX-0 laminate, the vector network analyser E8361A as the reader and the chipless tag. The experiment was conducted in a laboratory and not in the anechoic chamber in order to validate the results in a real-life reflective environment. The block diagram and photograph of the experimental setup are shown in Figs. 7.22 and 7.23, respectively.

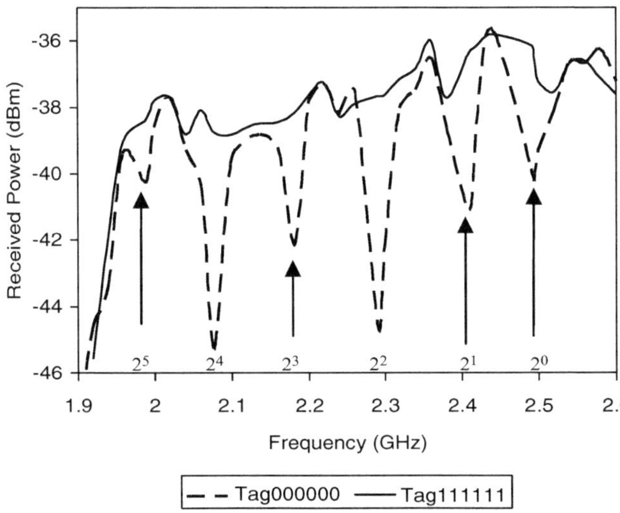

Fig. 7.24 Received power levels by reader for chipless tags at 5 cm

We encoded the tag with ID 000000 and placed it 5–25 cm from the reader antennas. The measured results of the received data are presented in both magnitude and phase with the variation of the tag-to-reader distance. Agilent's E8361A network analyser was used as the reader. The analyser was calibrated with the output power of the ports being −28 dBm. The variation of the received power magnitude due to the tag frequency signatures is presented in Fig. 7.24 and Table 7.2.

We can see that the tag's readability goes down with the reading distance. The tag was successfully read at distances up to 15 cm. However, the reading distances of 20 and 25 cm introduced reading errors at 2 and 2.4 GHz. The reading errors show that the spectral signature encoded into the magnitude is subject to noise and interference and can result in reading errors. Hence, we have introduced data encoding into the phase of the spectral signature, which should increase the reading range. The phase information in the spectral signature was measured and assessed in such a way that the tag with the 111111 ID is used as a reference. The phase variation with frequency of the received tag signals is presented in Fig. 7.25 and Table 7.2.

From Table 7.2, it is clear that the tag is accurately read by the reader (even at 25 cm). Thus, we can conclude that the phase information is more resilient to noise and can be read from a greater distance when compared to the amplitude information of the frequency signature. This represents a novelty in data encoding and data extraction when it comes to spectral signature-based chipless RFID systems.

The testing of the Gen-2 RFID (amplitude and phase detection) reader (2–2.5 GHz) was conducted in an anechoic chamber using log periodic dipole (LPDA) arrays operating between 1.9 and 2.7 GHz as the reader antennas, as shown in Fig 7.23. The Gen-2 RFID reader was tested without the testing of the Gen-1 since it provided an integrated solution of both amplitude and phase detection. The 6-bit chipless RFID tag was placed up to 10 cm from the reader antennas, but in this

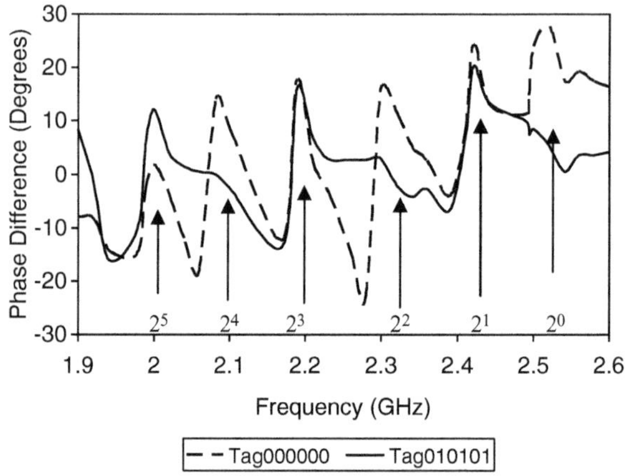

Fig. 7.25 Phase variations of the received tag signal at the reader end for chipless tags at 5 cm

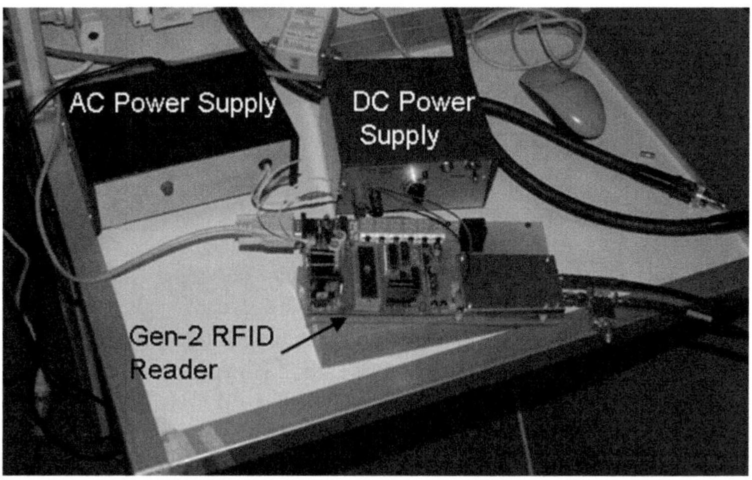

Fig. 7.26 Photograph of RFID reader in operating mode

case the tag is placed in between the two reader antennas in order to minimize mutual coupling and cross-talk between the reader antennas. The RFID reader in operating mode is shown in Fig. 7.26.

The RFID reader interrogates the tag by sweeping the frequency spectrum in 150–180 samples (amplitude data in 180 points and phase data in 150 points). The flow chart of the algorithm is presented in Fig. 7.11. It was necessary to calibrate the RFID reader first by setting up the experiment and then determining the necessary thresholds of the amplitude and phase data, which determined logics “1” and “0”. This was done by interrogating a tag with all logic zeros in its ID and recording the data and then replacing the tag with all logic ones in its ID. Hence, a clear difference between what is logic “0” and what is logic “1” is created and recorded in the reader. The calibration procedure enables the detection of a tag in the reader’s interrogation area by measuring the received signal strength. The tag spectral signatures (“000000”

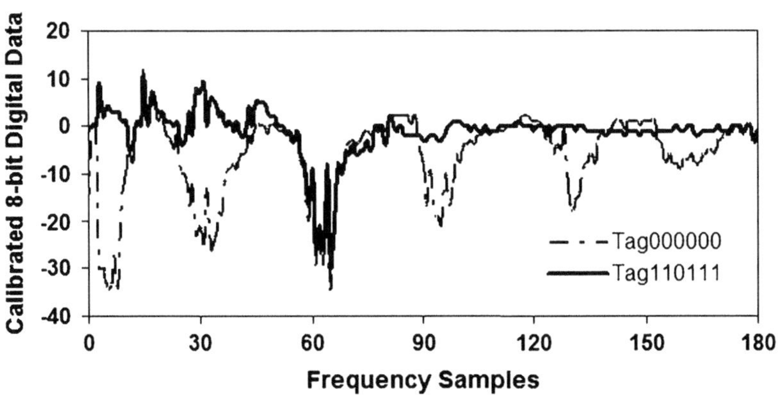

Fig. 7.27 Recorded amplitude of tag's spectral signature after calibration by RFID reader prototype

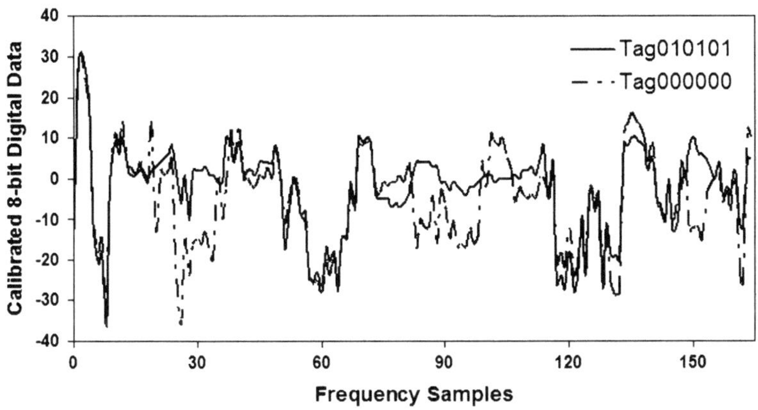

Fig. 7.28 Recorded phase of tag's spectral signature after calibration by RFID reader prototype

and "110111") calibrated by the reader in both amplitude and phase are presented in Figs. 7.27 and 7.28, respectively.

The tag ID is decoded by using the threshold values established in the calibration routine. After the tag ID has been decoded, it is sent to the display and/or via RS-232 to the host computer application. The algorithm returns to "Tag Interrogation" (Fig. 7.11) mode and waits to detect the presence of a tag.

The successful reading of the 6-bit proof-of-concept tag by the PNA at distances up to 40 cm in the anechoic chamber and 25 cm in real-life environment has confirmed that the chipless RFID system is more than suited for the polymer banknote tagging application. The developed Gen-2 reader could detect the tag at the minimum set distance of 10 cm successfully. The PNA read the tag at a further reading range since its sensitivity and architecture is more advanced than the Gen-2.

The following section presents the field trials of the UWB chipless RFID system.

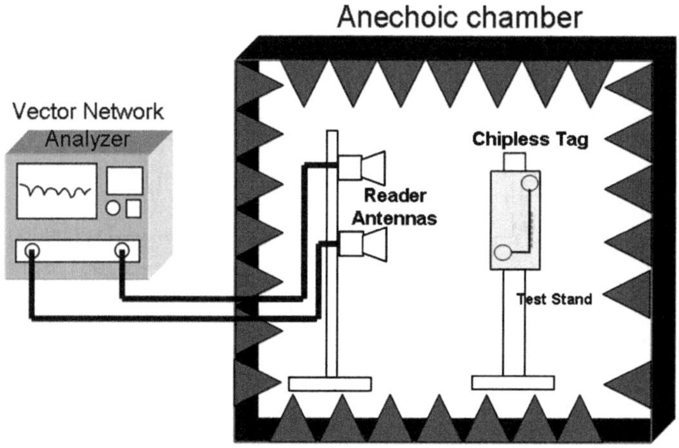

Fig. 7.29 Block diagram of the anechoic chamber setup

7.4.2 UWB RFID Chipless Tag: Reader System Field Trials

The experimental setup in the anechoic chamber is composed of the chipless tag, the vector network analyser (VNA) PNA E8361A as the reader electronics and horn antennas as the reader antennas. Horn antennas were used to increase the reading range of the tag since they have high gain (~11 dBi). The experiment was conducted in the Monash University Anechoic Chamber in order to validate the successful encoding of the tag and its detection at the reader end using the network analyser. The chipless tag and the reader antennas were mounted on plastic stands and placed into the anechoic chamber. A block diagram of the experimental setup is shown in Fig. 7.29.

As the horn antennas covered the frequency range from 7 to 12 GHz hence the tag was interrogated starting from 7 GHz. This resulted in reading 13 bits of the entire 23-bit data encoded by the tag. However, this was sufficient to prove the successful operation of the tag and provide a read range estimation using horn antennas. A photograph of the experimental setup is shown in Fig. 7.30.

The use of horn antennas as reader antennas in this experimental setup greatly increased the reading range of the tag. We attribute this to the greater isolation the cross-polarized reader antennas, higher directivity and higher gain than those of the log periodic arrays (presented in Chap. 4). The cross-polar reader antennas are shown in Fig. 7.31. As can be seen from Fig. 7.32, the isolation between the reader antennas is well above 65 dB.

We encoded the tag with ID "0x000000" and placed it from 5 to 70 cm (in steps of 5 cm) away from the horn reader antennas as shown in Fig. 7.30. The PNA was calibrated with the output power at the ports being –28 dBm. Both amplitude and phase data were retrieved when interrogating the tag. The chipless RFID tag was detected using a reference tag "0x111111", which carried no resonances. Hence, when the two results were compared, the encoded resonances from tag ID

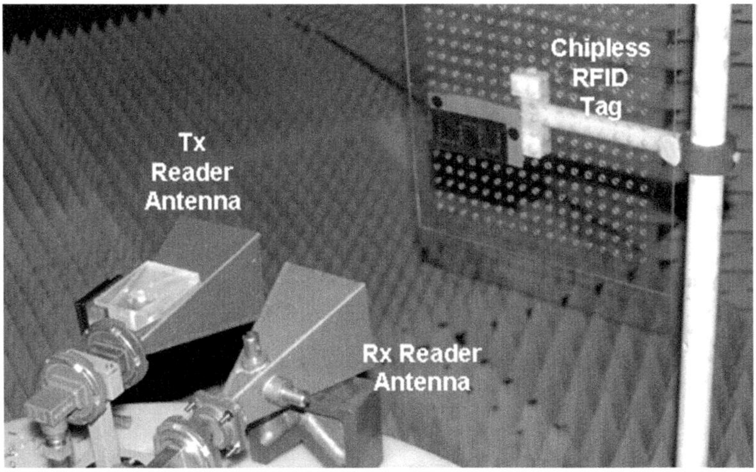

Fig. 7.30 Photograph of the experimental setup in the anechoic chamber of UWB RFID system

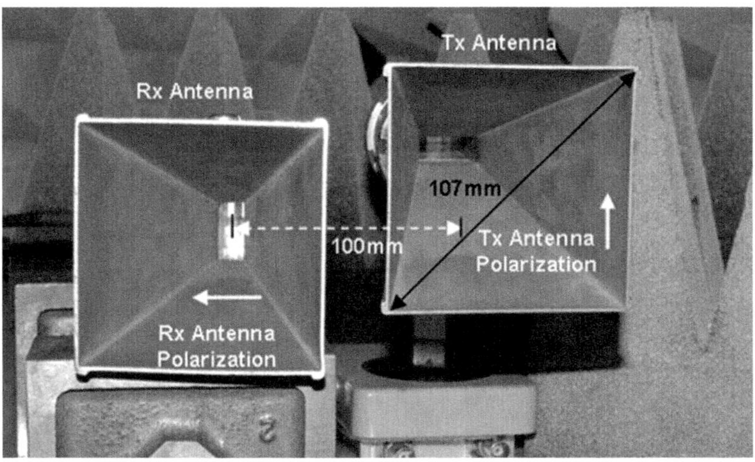

Fig. 7.31 Photograph of cross-polarized horn antennas used at reader end with 10 cm separation

"0x000000" were successfully detected. The normalized magnitude and phase of tag ID "0x000000" at 10 cm are presented in Figs. 7.33 and 7.34, respectively. The measured results vs. distance of tag from reader antennas are shown in Fig. 7.35.

From Fig. 7.35, it is clear that in the anechoic chamber the tag can be detected further away (up to 70 cm) when using phase data detection than when using amplitude data detection. This is attributed to the greater robustness of phase when compared to amplitude. The successful interrogation of the tag in both amplitude and phase was conducted up to 50 cm. This result shows an improvement in the reading range detection of 300% in amplitude data and 75% in phase data (up to 70 cm)

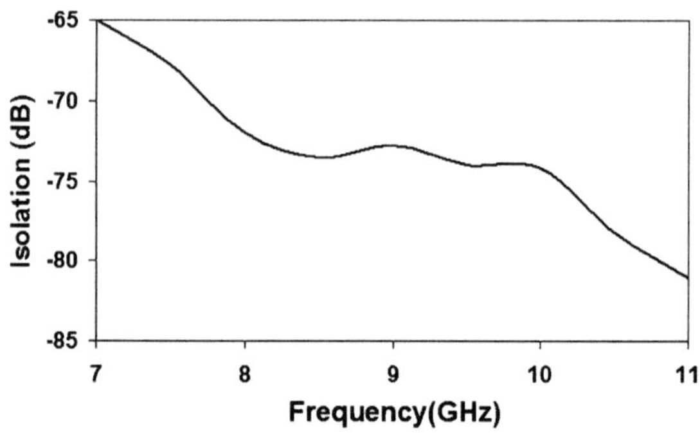

Fig. 7.32 Measured isolation between cross-polarized reader horn antennas

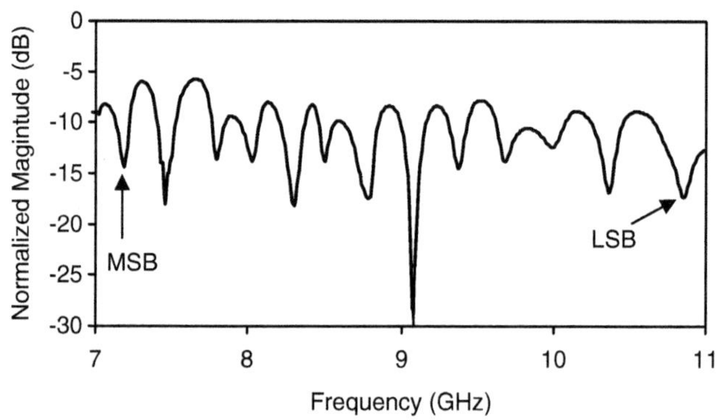

Fig. 7.33 Normalized magnitude variation vs. frequency of chipless RFID tag with ID "0000000000000" from 7 to 10.7 GHz

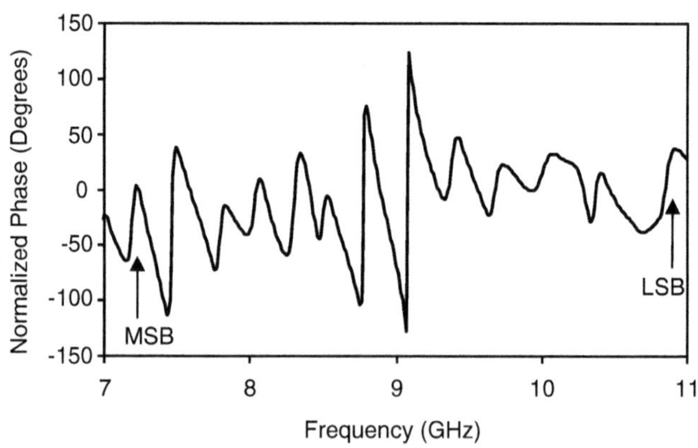

Fig. 7.34 Normalized phase variation vs. frequency of chipless RFID tag with ID "0000000000000" from 7 to 10.7 GHz

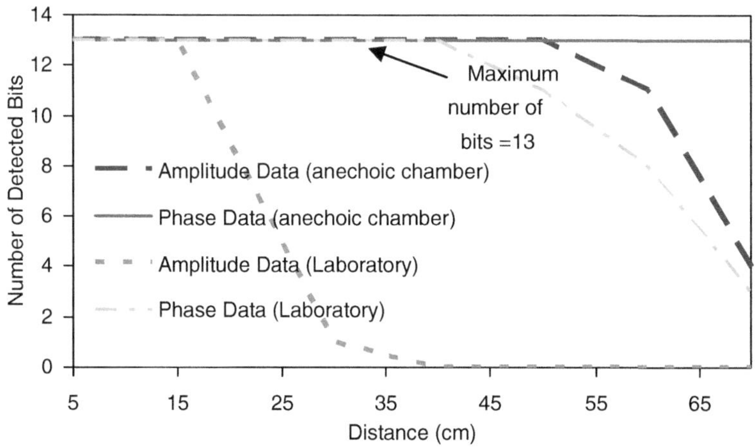

Fig. 7.35 Number of successfully detected bits vs. distance of tag from reader antennas from 7 to 10.7 GHz (maximum of 13 detectible bits)

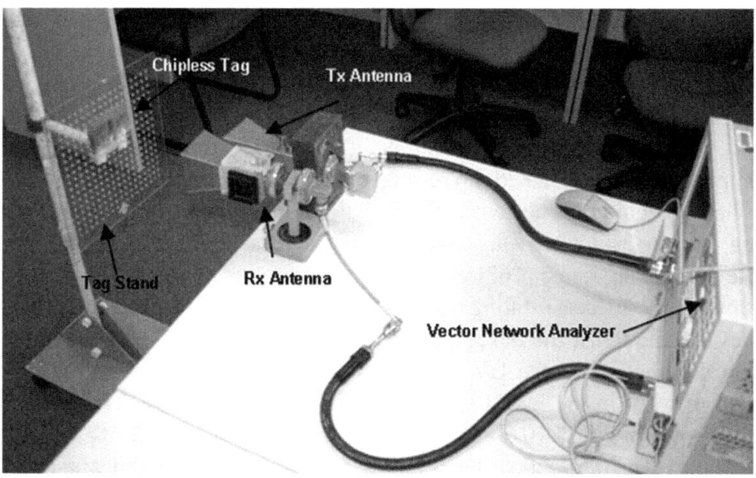

Fig. 7.36 Photograph of the experimental setup in the laboratory

compared with results reported in the previous section. The increased reading range in amplitude was greatly influenced by the increase of the cross-polar isolation of the tag antennas, increased isolation between the reader horn antennas and higher gain of the reader antennas (~11 dBi over the entire band).

The chipless tag was placed in a laboratory setup (outside the anechoic chamber Fig. 7.36) in order to measure the detection range of this particular setup when exposed to environmental influences.

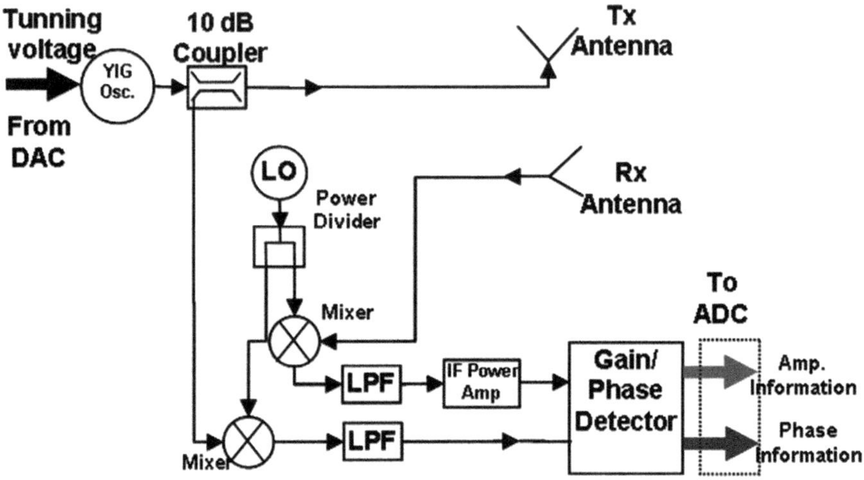

Fig. 7.37 Block diagram of second Gen UWB RF section

Figure 7.35 shows that the tag was read accurately in both amplitude and phase up to 15 cm when placed in a laboratory as shown in Fig. 7.36. The phase data was detectable at greater reading ranges (up to 35 cm) than the amplitude data due to robustness of the phase data. Figure 7.35 clearly shows that the reading range dropped by 50% outside the anechoic chamber due to interference from the environment. Although, it should be mentioned that the detection procedure was a simple comparison of tag data with no resonances and tag data with all resonances. The reading range could be improved by using signal processing techniques (matched filtering, etc.) to isolate the tag signal from the noise and interference, and thus increase the reading range [163].

Following the successful long range reading of the tag under test using two horn antennas and the PNA, the developed UWB RFID reader was installed in the system. The RF section of the UWB reader operating between 1.9 and 2.5 GHz is upgraded to work in the UWB spectrum as shown in Fig. 7.37. The entire RFID reader is shown in Fig. 7.10 composed of the upgraded digital and RF sections and horn antennas. The digital board was upgraded with an LCD display and two 10-bit ADC instead of one 8-bit ADC so that both amplitude and phase data could be sampled simultaneously at a higher resolution.

The UWB transceiver consists of a transmitter and receiver path. The transmitter is in the form of a Teledyne YIG oscillator, which generates an interrogation signal of constant 15 dBm power to the Tx antenna. The RFID reader interrogates the tag by sweeping the RF signal from 7 to 10.7 GHz. The readings are performed from 7 to 10.7 GHz due to the cutoff frequency of the horn antennas at 7 GHz. This results in reading a tag of 13 bits from 7 to 10.7 GHz by the UWB RFID reader.

The UWB chipless RFID system experimental setup is shown in Fig. 7.36. In this experiment, the reader performed wireless interrogation and detection of the tag. The same digital section and decoding algorithm, as shown in Fig. 7.11, was used to interrogate, detect and decode the tag. The 10-bit digital amplitude and

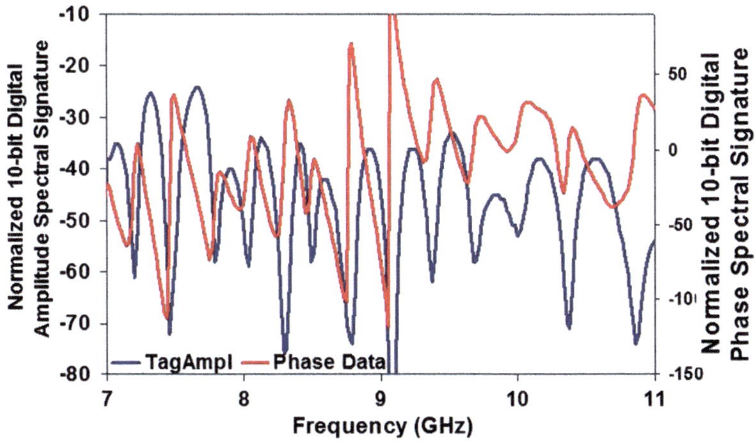

Fig. 7.38 Normalized 8-bit digitized tag amplitude and phase spectral signature obtained from wireless readings using UWB RFID reader prototype

Table 7.3 Amplitude and phase differences vs. frequency between Tag000000 and Tag111111 (ΔA—amplitude difference in 10-bit digital data and $\Delta\theta$—phase difference in 10-bit digital data)

Frequency (GHz)	Data bit	Amplitude difference	Phase difference
7.18	2^{12}	60	59
7.46	2^{11}	76	140
7.8	2^{10}	58	59
8.04	2^{9}	58	28
8.3	2^{8}	77	24
8.5	2^{7}	59	30
8.76	2^{6}	74	170
9.1	2^{5}	94	192
9.38	2^{4}	62	55
9.7	2^{3}	58	41
10.02	2^{2}	52	32
10.38	2^{1}	71	42
10.7	2^{0}	74	70

phase spectral signatures of the tag are shown in Fig. 7.38. The received 13-bit tag signal is normalized in respect to a tag with no resonances. The 13 amplitude nulls and phase jumps of tag data are detected between 7 and 10.7 GHz as expected. Table 7.3 shows the amplitude and phase difference between the reference tag (all ones) and the tested tag with all resonances present (all zeros). From Table 7.3, it is clear that the 10-bit digital data difference is large enough to be distinguished by the RFID reader's threshold detection algorithm presented in Fig. 7.11. The tag was detected up to 15 cm using the UWB RFID reader. The significance of this successful reading using the UWB RFID reader is that it is the first UWB RFID reader which has been developed to detect the chipless RFID tag spectral signature in both amplitude and phase.

With the successful detection of the tag by the UWB RFID reader at distances greater than 10 cm the field trials of the chipless RFID tag-reader system are finalized. The chipless RFID system has been successfully tested inside and outside the anechoic chamber and yielded satisfactory results by detecting the chipless tag using an in-house built RFID reader.

7.5 Conclusion

In this chapter, we have presented a novel fully printable chipless RFID system with field trials which can be used for tracking low-cost items, such as note bills, envelopes and other paper/plastic products, items and documents. The chipless RFID tags use multiresonators to encode data into the spectral signature. By interrogating the tag by a multi frequency signal, it is possible to detect the variations in the magnitude and phase of the received tag signal and decode the tags ID.

Firstly, the proof-of-concept chipless RFID system was tested. The chipless RFID tag was first interrogated using Agilents PNA E8361A inside an anechoic chamber. The results yielded a reading range of up to 40 cm using phase data of the spectral signature, while accurate readings of the tag using amplitude spectral signature were conducted up to 10 cm. Field trials were then set outside the anechoic chamber where the influences of reflections and multipathing reduced the reading range of the system up to 25 cm. Finally, the 6-bit proof-of-concept chipless RFID system was completed using the Gen-2 RFID reader instead of the PNA for amplitude and phase spectral signature detection. The chipless tag was successfully detected at 10 cm using the Gen-2 RFID reader, which was the minimum required reading range set.

Future works will be concentrated on better accuracy of reading, developing the chipless system on plastic substrates and RFID reader development with anticollision protocols.

The UWB chipless RFID system which utilizes a fully printable chipless CPW RFID tag which can be used for tracking low-cost items, such as banknotes, envelopes and other paper/plastic products, items and documents, and UWB RFID reader were tested. The chipless RFID tag operates between 5 and 10.7 GHz of the UWB spectrum. By exciting the tag with a wideband signal, it was possible to detect variations in the magnitude and phase of the received tag signal and decode the tag's ID at distances up to 70 cm in a noise-free environment and up to 35 cm in a laboratory (noisy) environment. It was necessary to calibrate the reader with a reference signature ID with no resonances when performing amplitude and phase data decoding.

Two prototype UWB RFID readers operating successfully between 5 and 10.7 GHz have been presented. The reader is a short range prototype. The lower frequency range RFID reader has successfully detected and decoded a 6-bit chipless tag at 10 cm using a software code, which utilizes calibration, threshold detection and amplitude and phase data decoding for tag ID extraction. A UWB reader with upgraded RF and digital section and same software algorithm detected a 13-bit tag. Wired readings of the tag using the UWB reader showed that it is possible to detect all 23 bits of the tag if an antenna operating between 5 and 10.7 GHz was used. The UWB RFID reader is the first UWB RFID reader to be built which can wirelessly interrogated and detect a chipless RFID tag using amplitude and phase spectral signature detection. The reader relies on calibration in order to provide accurate readings of the tag. The RFID reader can be mounted on conveyor belt applications.

The spectral signature-based chipless RFID system is the first of its kind to use phase to encode data, which increases the reading range and accuracy and provides the reader with the ability to interrogate magnitude and phase and compare them to confirm the ID of the interrogated tag. It is necessary to use a reference signature ID of 111111 when performing amplitude and phase data decoding.

The summary of the achievements that have been accomplished while developing and testing the chipless RFID tag and RFID reader and their integration into a chipless RFID system is as follows:

- Phase data of the chipless tag enhances the detection performance and reading range.
- A thorough investigation of the RFID system properties, such as tag and reader antenna cross-talk, amplitude and phase tag data extraction, robustness of data processing and data display.
- Integrated and compact Gen-1 and Gen-2 reader modules which can be handheld as well.
- An RFID reader calibration process for data decoding based on using a reference tag with no resonances (all ones). The calibration reduces the impact of environmental interference and system noise.
- Interference free (anechoic chamber) and laboratory tests to show the system functionality and reader performance.
- A novel approach of utilizing the UWB spectrum for more tag data bits.
- Flexible CPW tag on thin laminates which can be transferred to plastic and/or paper and its successful detection by the RFID reader.
- Integration of multidisciplinary engineering fields such as microwave engineering, antenna design, digital circuit design, application software and reader algorithm programming into an RFID system level design.
- Successful reading ranges in the anechoic chamber of up to 70 cm is a unique achievement for a fully passive chipless RFID system.
- Successful field trials of the chipless tags readers place the development in the forefront of the chipless RFID world.

The system has a great potential if printed on plastic substrate using transparent conductive (silver) ink for low-cost item tracking. It would represent a cheap and economical way of potentially replacing the barcode due to the fact that the chipless tags are not comprised of any silicon-based circuits or semiconductor devices.

Chapter 8
Conclusions and Future Works

Abstract In this chapter, we present the conclusions and proposed the future works that can be conducted in order to improve the performance of the chipless RFID tag and RFID readers used to detect the tag.

The content of this book has been concerned with the design of chipless RFID tags and RFID readers for low cost item tracking. The growing tendency today is to replace the barcodes with RFID tags. Knowing the limitation of the optical barcode, RFID provides unique ID codes for individual items that can be read at a longer distance. Hence, the obstacles of reading range and automation would be addressed by the use of RFID. However, the cost of RFID tags makes them unaffordable as an alternative to barcodes.

In recent years, chipless RFID has been proposed as a low cost and competitive replacement for the barcode. The recently reported chipless RFID tags are printed resonators [59], chemical fibres [53] and TFTC [49] organic tags. However, these reported chipless tags have been stagnating in the prototyping stage and have limitations in terms of reading range, size, data capacity, data encoding, frequency of operation and finally, fabrication challenges. As for an example, printed resonators have size restrictions, chemical fibres have reading ranges up to a couple of millimetres, while TFTC has very low electron mobility and can only operate in the kHz range and at best MHz frequency range.

Another aspect of chipless RFID system development, which has been ignored so far by researchers, is the design of RFID readers for chipless tags. Since the chipless RFID tags use unconventional methods for data encoding, such as spectral signatures, conventional "off-the-shelf" RFID readers are not suitable for the new development. Hence, the RFID readers were needed to be developed from scratch.

S. Preradovic and N.C. Karmakar, *Multiresonator-Based Chipless RFID: Barcode of the Future*, DOI 10.1007/978-1-4614-2095-8_8,

In response to this situation, the research/design work accomplished in this book has concentrated on the design and development of low cost, fully printable chipless RFID tags and dedicated RFID readers. The chipless RFID system is to be used for tagging the Australian polymer banknote in a conveyor belt setup. Two varieties of chipless RFID tags, on conventional Taconic TLX-0 laminate and thin flexible Taconic TF-290 laminate, have been designed and tested. A chipless RFID tag—reader system which can detect the tag has been presented.

8.1 Accomplished Designs and Outcomes

As the main goals of this project, fully printable chipless RFID tags on conventional PCB and thin flexible laminates, which encode data using spectral signatures, and dedicated RFID readers, which are used to detect chipless RFID tags, have been developed. The chipless RFID system, comprising the chipless tags and RFID readers, has been fully tested and is designed for tagging the Australian polymer banknotes.

The development and successful testing of the chipless RFID system answer the demand for a fully printable ultra-low-cost tag and an RFID reader, which can detect these tags while mounted over a conveyor belt. The salient feature of the novel chipless RFID tag is its fully printable single-layered design in a compact and low cost format. It has significant amount of data-encoding capability (up to 35 bits were designed). The dedicated chipless tag RFID reader is a novel mixed signal circuit. The salient features of the developed chipless tag RFID reader are its UWB operation and capability to decode frequency signatures into a 1:1 correspondence of binary data bits without requirements of any handshaking algorithm between the reader and the tag.

Three main tasks had been identified and set at the start of this book: (1) chipless RFID tag development, (2) chipless tag RFID reader development and (3) chipless RFID system integration. All three tasks had been successfully completed and presented in the preceding chapters of this book.

Prior to the design and development of the chipless RFID tag and reader devices, a comprehensive literature review of RFID tags and readers was conducted. The goal of the literature review was to identify the niche areas of design and development in RFID, where novel research could be carried out. The comprehensive literature review on chipless RFID tags and readers revealed that chipless tags which are fully printable, multi-bit with ease of data encoding were not available. Some work had been conducted on capacitively tuned dipoles [59] and fractal Hilbert curve [60] based tags, but without the ability of data encoding. The comprehensive review on RFID readers revealed that conventional RFID readers were dedicated for HF (13.56 MHz) and UHF (915 MHz) RFID chipped tags. These readers use ASK and PSK modulation techniques in contrary to the spectral signature modulation of the chipless RFID tag. These findings steered the research towards the development of a novel multiresonator-based chipless RFID tag and dedicated RFID reader both operating over the UWB frequency band. The review was concluded with the proposed chipless RFID tag and RFID reader and their design specifications.

Based on the tag and reader specifications, it was clear that the chipless RFID tag comprises two main components: UWB antenna and multiresonator. The multiresonating circuit consisted of cascaded spiral resonators, which operate at different resonant frequencies. Each resonant frequency corresponded to a single data bit. The spiral resonator was chosen as the main encoding element since it exhibit compact size, high Q and small bandwidth in comparison with other planar resonators, which exhibited stop-band performance. A comprehensive theoretical and parametric study of the spiral resonator on microstrip technology using standard PCB laminate (Taconic TLX-0) resulted in the optimal layout parameters of the spiral resonator. Following the optimized design of the spiral resonator on PCB, the spiral resonator was designed on thin 90 μm flexible laminate (Taconic TF-290). The microstrip spiral resonator exhibited a dramatic drop in Q factor, which was completely unacceptable for implementation in the chipless tag. The reduction in Q factor was due to the increase in dielectric loss. The substrate thickness was electrically very small in comparison with the wavelength of the propagating EM signal and increased metallic losses, which reduced the coupling between the spiral resonator and microstrip line.

The use of CPW spiral resonators on thin flexible laminate TF-290 solved the issue of low Q factor. A comprehensive parametric study of the CPW spiral resonator was conducted in order to optimize the performance of the spiral resonator. Following the successful optimization of the CPW spiral resonator, the spiral resonators were cascaded to form a multiresonating circuit. The multiresonator is used to encode the unique spectral signature of the chipless RFID tag. A parametric study of cascaded spiral resonators has shown that mutual coupling between spirals is minimal for spiral resonator separation of 3 mm.

Spectral signature encoding is used to encode data by the tag. Spectral signature requires a one-to-one (1:1) correspondence of the frequency spectrum behaviour to the tag's multiresonator layout. In particular, each spiral resonator had a 1:1 correspondence with a data bit, which meant that each data bit had a predetermined spiral resonant frequency. To the best of the candidates' knowledge, spectral signature encoding utilizing both amplitude and phase of the spectral signature is the first of its kind and was not reported before. The spiral resonance was represented by a null in the amplitude and abrupt jump in the phase, which encoded logic "0". Encoding logic "1" was represented by the absence of an amplitude null and phase jump.

A fully novel "spiral shorting" concept of data encoding is presented in this book. The spiral resonator is shorted by shorting the spiral turns with a single trace. When shorted, the spiral resonator has a resonant frequency, which is outside the operating band of the chipless RFID tag hence resulting in the absence of the resonance. This is characterized as a logic "1" bit in the spectral signature. Removing the shorting between the spiral turns introduces the resonance of the spiral resonator, which is a representation of logic "0". This novel data-encoding technique has given new manufacturing advantage of the chipless RFID technology over other reported chipless RFID tags in terms of minimum layout modifications and the use of laser etching for mass tag encoding.

Following the successful design of the multiresonator, the design of the UWB monopole antennas for the chipless RFID tags was performed. The UWB disc-loaded monopole antennas exhibit omni-directional radiation patterns over their operating band and have an efficient and compact layout. The monopoles were designed using microstrip and CPW technology. CPW proved to have an advantage over microstrip since the monopole's layout was single layered and hence was easy to print.

The RFID reader antennas that were designed in this book were log periodic dipole antennas (LPDAs). They exhibit wide bands of operation, high gain and directional radiation patterns. The use of directional high gain antennas, as reader antennas, has been justified as to provide spatial diversity of the interrogation zone through which the chipless tags move on a conveyor belt.

In Chap. 5, the integration of the UWB monopole antennas (Chap. 4) and multi-resonators (Chap. 3) has been performed to form the chipless RFID tag. Three chipless tags were designed: (1) a 6-bit proof-of-concept chipless RFID tag operating from 2 to 2.5 GHz, (2) a 35-bit microstrip UWB chipless RFID tag on conventional PCB laminate and (3) a 23-bit CPW UWB chipless RFID tag on a flexible thin laminate. The proof-of-concept tag validated the concept of the novel chipless RFID tag. The 35- and 23-bit tags showed that multiple data bits were encoded using the novel chipless tag concept. The CPW 23-bit chipless RFID tag showed that the design is transferable from laminate to plastic since the TF-290 has extremely similar electrical characteristics as polymer substrates used for the Australian polymer banknote. The chipless tags were wirelessly interrogated inside an anechoic chamber and were successfully detected using an Agilent PNA as the reader electronics. Both amplitude and phase of the spectral signature were detected. The successful reading of the chipless tags confirmed the viability of the chipless RFID system.

Following the successful design of the chipless RFID tags, the design of a chipless RFID reader was performed. The architecture of an RFID reader is composed of the reader antennas (LPDAs), RF transceiver electronics and a digital section. Three different RF transceiver architectures were designed for the dedicated chipless tag RFID readers: (1) Gen-1, (2) Gen-2 and (3) UWB transceivers. The Gen-1 and Gen-2 RF transceivers operated between 2 and 2.5 GHz and were designed to interrogate the 6-bit proof-of-concept RFID tag. The Gen-1 transceiver was the first circuit to be designed and only had amplitude detection capabilities of the spectral signature using a diode detector. The Gen-2 transceiver was upgraded using a gain/phase detector in the receiver, which was used to detect the amplitude and phase of the tag's spectral signature. Finally, the UWB RF transceiver circuit is designed by upgrading the Gen-2 transceiver to operate in the UWB region. Since the gain/phase detector circuit operated up to 2.7 GHz, downconversion mixers and local oscillator were implemented to downconvert the received signal from UWB to below 2.7 GHz. All three transceiver architectures were tested using a wired tag. The wired test was performed in order to characterize the performance of the RF transceivers. The RF transceivers were characterized by measuring the transmitter output power and higher order mode distortion, isolation between the transmitter and receiver and receiver sensitivity.

All three transceiver circuits met the predetermined characteristics, which meant that the system level integration was ready to be performed.

Finally, the chipless RFID tags and RFID readers were integrated into a chipless RFID system. The frequency signature-based system was the first of its kind to use phase to encode data with increased reading range. The Gen-1 and Gen-2 prototype RFID readers operated successfully between 1.9 and 2.5 GHz, and the UWB RFID reader operating between 7 and 10.7 GHz was used to detect the chipless tags. The readers were short range prototypes and are proof-of-concept circuits/devices. The lower frequency range RFID readers, Gen-1 and Gen-2, had successfully detected and decoded a 6-bit chipless tag at 10 cm using a software code, which utilizes calibration, threshold detection and amplitude and phase data decoding for tag ID extraction. A UWB reader with upgraded RF and digital section and same software algorithm detected a 13-bit tag at 15 cm. The reading range of the system was tested using an Agilent PNA as the reader. By exciting the tag with a wideband signal, it was possible to detect variations in the magnitude and phase of the received tag signal and decode the tag's ID at distances up to 70 cm in a noise-free environment and up to 35 cm in a laboratory (noisy) environment. The UWB RFID reader was the first UWB RFID reader to be built for wireless interrogated and detect of chipless RFID tags using amplitude and phase spectral signature detection. The reader relied on calibration in order to provide accurate readings of the tag. The RFID reader can be mounted on conveyor belt applications.

From the results presented in this book, we conclude that the research and design objectives were fulfilled. The chipless RFID system presents an ultra-low-cost solution for low cost item tagging. In summary, the project has resulted in many significant contributions to the areas of RFID tag design, antenna design, mixed signal circuit design and microwave engineering in general. The research work presented in this book has generated much interest from industry and researchers both within Australia and overseas. Three large ARC Linkage Grants, which are worth more than AUD $2 million, frequent visits from the industry partners and academic researchers from Australia and overseas, two Australian and international patents and many peer-reviewed journal and conference proceedings papers have signified the claim of fulfilling the goals of the research.

8.2 Future Works and Open Issues

With the high demand on RFID technology in terms of reading range and applications, some open issues and further works remain to be addressed in future projects. So far, the RFID tag has been designed to operate in predefined alignment situations and applications, since the polarization of the antennas is crucial for successful readings. Further works could focus on developing planar circularly polarized tag antennas, which would remove the present stringent alignment requirements. Another improvement which could be considered is making the tag operate with a single antenna instead of two, which would dramatically reduce the size of the chipless tag.

Size reduction of the chipless tag can be further increased by using submillimetre- and millimetre-wave frequency bands. New applications for chipless tags (such as tram and train ticketing, etc.) could be established by extended the capacity of the chipless tags to 124 bits.

The RFID readers presented in this book are short range reader prototypes. System level testing using the PNA showed that the tag can be read up to 70 cm. Future works should focus on developing a long range RFID reader by using phased or switched beam antenna arrays as reader antennas and transceivers with higher receiver sensitivity. Novel applications of the chipless RFID system could be expanded by anticollision protocols implemented in the RFID reader. The basis of these anticollision protocols should be on spaced division multiple access (SDMA) and time division multiple access (TDMA) protocols.

Acknowledgements The project work has been fully supported by the prestigious Australian Research Council's (ARC) large Discovery Project grant DP0665523 "Chipless RFID for Barcode Replacement". The work associated with the design and development of this chipless RFID system has formed a substantial part of a large ARC grant project, which has been awarded to Dr. Nemai Chandra Karmakar.

References

1. K. Finkenzeller, "RFID Handbook – 2nd Edition", John Wiley & Sons, Ltd., 2003.
2. U. Kraiser, W. Steinhagen, "A low-power transponder IC for high-performance identification systems", *IEEE Journal of Solid-State Circuits*, vol. 30, no. 3, pp: 306–310, March 1995.
3. S. Preradovic, N. Karmakar, "Modern RFID readers", *Microwave Magazine,* internet article, Available: http://www.mwjournal.com/article.asp?HH_ID=AR_4830 (Accessed August 2009).
4. H. Stockman, "Communication by Means of Reflected Power", Proceedings of the IRE, pp: 1196–1204, October 1948.
5. B. O'Neill, "Canadian Wall-Marts implement RFID to eliminate out-of-stocks", AcessmyLibraty Flexible Packaging Internet Article, January 2008. Available: http://www.accessmylibrary.com/coms2/summary_0286-33829113_ITM (Accessed August 2009).
6. Transponder News, "EAS systems", Internet article, Dec 2000. Available: http://www.transpondernews.com/easbasic.html (Accessed September 2009).
7. Document Tracking Systems, "Using RFID to track Documents", Whitepaper of Intensecomp Pte Ltd., Dec 2000, Available: http://www.intensecomp.com/wtp_docutracking.html (Accessed September 2009).
8. National Livestock Identification System, official website, www.nlis.mla.com.au.
9. D'Hont S., The Cutting Edge of RFID Technology and Applications for Manufacturing and Distribution", Texas Instruments white paper, http://www.ti.com/rfid/docs/manuals/whtPapers/manuf_dist.pdf (accessed March 2006).
10. S. Preradovic, N. Karmakar, I. Balbin, "RFID Transponders", *IEEE Microwave Magazine*, vol. 9, no. 5, pp: 90–103, Oct. 2008.
11. P. Harrop, The price-sensitivity curve for RFID", IDTechEx internet article, August 2006. Available: http://www.idtechex.com/products/en/articles/00000488.asp (accessed May 2008).
12. J. Collins, "Alien cuts tag price", RFID Journal internet article, April 2004. Available: http://www.rfidjournal.com/article/articleview/857/1/1/ (accessed April 2008).
13. R. Das, "Chipless RFID – The end game", IDTechEx internet article, February 2006. Available: http://www.idtechex.com/products/en/articles/00000435.asp (Accessed: December 2006).
14. S Mukherjee, "Chipless radio frequency identification by remote measurement of complex impedance", Proc. 10th European Conference on Wireless Technology, pp: 249–252, Munich, Germany, Oct. 2007.
15. C. S. Hartmann, "A global SAW ID tag with large data capacity", Reprint form *Proceedings of 2002 IEEE Ultrasonics Symposium*, vol. 1, pp: 65–69, Munich, Germany, October, 2002, Available: http://www.rfsaw.com/pdfs/Global_SAW_ID_Tag_lg.pdf.
16. L. Zheng, S. Rodriguez, L. Zhang, S. Botao, L-R. Zheng, "Design and implementation of a fully reconfigurable chipless RFID tag using Inkjet printing," *IEEE International Symposium on Circuits and Systems ISCAS 2008*, pp: 1524–1527, Washington, USA, May 2008.

S. Preradovic and N.C. Karmakar, *Multiresonator-Based Chipless RFID: Barcode of the Future*, DOI 10.1007/978-1-4614-2095-8,

17. S. Shrestha, M. Balachandran, M. Agarwal, V. V. Phoha, K. Varahramyan, "A chipless RFID sensor system for cyber centric monitoring applications," *IEEE Transactions on Microwave Theory and Techniques*, vol. 57, no. 5, pp: 1303–1309, May 2009.
18. J. McVay, A. Hoorfar, N. Engheta, "Space-filling curve RFID tags", *2006 IEEE Radio and Wireless Symposium*, pp: 199–202, San Diego, USA, 17–19 Jan. 2006.
19. I. Jalaly, D. Robertson, "Capacitively-tuned split microstrip resonators for RFID barcodes", *2005 European Microwave Conference*, vol. 2 pp: 4, Paris, France, 4–6 Oct. 2005.
20. S. Preradovic, I. Balbin, N. C. Karmakar, G. F. Swiegers, "Multiresonator-based chipless RFID system for low-cost item tracking", *IEEE Transactions on Microwave Theory and Techniques,* vol. 57, no. 5, pp: 1411–1419, May 2009.
21. R. R. Fletcher, "Low-cost electromagnetic tagging: design and implementation", PhD Thesis, September 2002. Available: www.media.mit.edu/physics/publications/theses/97.02.fletcher.pdf.
22. D. A. Hodges, H. G. Jackson, "Analysis and design of digital integrated circuits – 2nd Edition", McGraw-Hill, New York, USA, 1988.
23. J. R. Baker, H. W. Li, D. E. Boyce, "CMOS circuit design, layout and simulation", IEEE Press, New York, USA, 1998.
24. S. Natarajan, "A 32nm logic technology featuring 2nd –generation high-k + metal-gate transistors, enhanced channel strain and 0.171μm² SRAM cell size in a 291Mb Array", *IEEE International Electron Devices Meeting 2008 IEDM 2008*, pp: 1–3, San Francisco, USA, 15–17 Dec. 2008.
25. S. Harma, V. P. Plessky, C. S. Hartmann, W. Steichen, "SAW RFID tag with reduced size", *IEEE Ultrasonics Symposium 2006*, pp: 2389–2392, Vancouver, Canada, Oct. 2006.
26. Y. Y. Chen, T. T. Wu, K. T. Chang, "A COM analysis of SAW tags operating at harmonic frequencies", *IEEE Ultrasonics Symposium 2007*, pp: 2347–2350, New York, Oct. 2007.
27. S. Harma, V. P. Plessky, L. Xianyi, P. Hartogh, "Feasibility of ultra-wideband SAW RFID tags meeting FCC rules", *IEEE Transactions on Ultrasonics, Ferroelectrics and Frequency Control*, vol. 56, no. 4, pp: 812–820, April 2009.
28. S. Harma, V. P. Plessky, X. Li, "Feasibility of ultra-wideband SAW tags", *IEEE Ultrasonics Symposium 2008*, pp: 1944–1947, Beijing, China, Nov. 2008.
29. P. Brown, P. Hartmann, A. Schellhase, A. Powers, T. Brown, C. Hartmann, D. Gaines, "Asset tracking on the international space station using global SAW tag RFID technology", *IEEE Ultrasonics Symposium 2007*, pp: 72–75, New York, Oct. 2007.
30. V. P. Plessky, S. N. Kondratiev, R. Stierlin, F. Nyffeler, "Saw tags: new ideas", *IEEE Ultrasonics Symposium 1995*, vol. 1, pp: 117–120, Cannes, France, Nov. 1995.
31. T. Han, W. Wang, H. Wu, Y. Shui, "Reflection and scattering characteristics of reflectors in SAW tags", *IEEE Transactions on Ultrasonics, Ferroelectrics and Frequency Control*, vol. 55, no. 6, pp: 1387–1390, June 2008.
32. T. Han, W. Wang, J. M. Lin, H. Wu, H. Wang, Y. Shui, "Phases of carrier wave in a SAW identification tags", *IEEE Ultrasonics Symposium 2007*, pp: 1669–1672, New York, Oct. 2007.
33. N. Saldanha, D. C. malocha, "Design parameters for SAW multi-tone frequency coded reflectors", *IEEE Ultrasonics Symposium 2007*, pp: 2087–2090, New York, Oct. 2007.
34. S. Harma, C. Kim, S. M. Balashov, V. P. Plessky, "Properties of narrow metal reflectors used in reflective array compressors and surface acoustic wave tags", *IEEE/MTT-S International Microwave Symposium 2007*, pp: 2051–2054, Honolulu, Hawaii, June 2007.
35. D. Puccio, D. Malocha, N. Saldanha, "Implementation of orthogonal frequency coded SAW devices using apodized reflectors", *IEEE International Frequency Control Symposium and Exposition 2005*, pp: 892–896, Vancouver, Canada, Aug. 2005.
36. S. Harma, W. G. Arthur, C. S. Hartmann, R. G. Maev, V. P. Plessky, "Inline SAW RFID tag using time position and phase encoding", *IEEE Transactions on Ultrasonics, Ferroelectrics and Frequency Control*, vol. 55, no. 8, pp: 1840–1846, August 2008.
37. S. Harma, V. P. Plessky, C. S. Hartmann, W. Steichen, "Z-path SAW RFID tag", *IEEE Transactions on Ultrasonics, Ferroelectrics and Frequency Control*, vol. 55, no. 1, pp: 208–213. January 2008.

38. S. Harma, W. G. Arthur, R. G. Maev, C. S. Hartmann, V.P. Plessky, "Inline SAW RFID tag using time position and phase encoding", *IEEE Ultrasonics Symposium 2007*, pp: 1239–1242, New York, Oct. 2007.
39. J. Liu, J. Yao, "Wireless RF identification system based on SAW", *IEEE Transactions on Industrial Electronics*, vol. 55, no. 2, pp: 958–961, Feb. 2008.
40. N. Saldanha, D. C. Malocha, "Low loss SAW RF ID tags for space applications", *IEEE Ultrasonics Symposium 2008*, pp: 292–295, Beijing, China, Nov. 2008.
41. L. Wei, H. Tao, S. Yongan, "Surface acoustic wave based radio frequency identification tags", *IEEE International Conference on e-Business Engineering 2008*, pp: 563–567, Xi'An, China, Oct. 2008.
42. J. Yao, J. Liu, B. Gao, "The realization of a passive identification tag transmitter based on SAW", *4th International Wireless Communications, Networking and Mobile Computing 2008*, pp: 1–4, Dalian, China, Oct. 2008.
43. J. Liu, J. Yao, "A new wireless RF identification system", *The Sixth World Congress on Intelligent Control and Automation 2006*, pp: 5191–5195, Dalian, China, June 2006.
44. S. Schuster, S. Scheiblhofer, A. Stelzer, A. Springer, "Model based wireless SAW tag temperature measurement", *Asia-Pacific Microwave Conference APMC 2005*, pp: 4–7, Suzhou, China, Dec. 2005.
45. E. L. Tan, Y. W. M. Chia, "Green's function and network analysis of quasi-2D SAW ID-tags", *IEEE Ultrasonics Symposium 2000*, pp: 55–58, San Juan, Puerto Rico, Oct. 2000.
46. D. Enguang, F. Guanping, "Passive and remote sensing based upon surface acoustic wave in special environments", *Microwave and Optoelectronics Conference 1997*, vol. 1. pp: 133–139, Natal, Brazil, Aug. 1997.
47. D. Puccio, D. C. Malocha, N. Saldanha, D. R. Gallagher, J. H. Hines, "Orthogonal frequency coding for SAW tagging and sensors", *IEEE Transactions on Ultrasonics, Ferroelectrics and Frequency Control*, vol. 53, no. 2, pp: 377–384, Feb. 2006.
48. G. Buckner, R. Fachberger, "SAW ID tag for industrial application with large data capacity and anticollision capability", *IEEE Ultrasonics Symposium 2008*, pp: 300–303, Beijing, China, Nov. 2008.
49. R. Das, P. Harrop, Chip-less RFID forecasts, technologies & players 2006 – 2016, IDTechEx internet article, Feb. 2006. http://www.idtechex.com/products/en/view.asp?productcategoryid= 96 (accessed March 2006).
50. A. Chamarti, K. Varahramayan, "Transmission delay line-based ID generation circuit for RFID applications", *IEEE Microw. Wireless Compon. Letters*, vol. 16, no. 11, pp: 588–590, Nov. 2006.
51. J. Vemagiri, A. Chamarti, M. Agarwal, K. Varahramyan, "Transmission line delay-based radio frequency identification (RFID) tag", *Microw. Opt. technol. Lett.*, vol. 49, no. 8, pp: 1900–1904, 2007.
52. S. Shrestha, J. Vemagiri, M. Agarwal and K. Varahramyan, "Transmission line reflection and delay-based ID generation scheme for RFID and other applications", *Int. J. Radio Freq. Identification Technol. Appl.*, vol. 1, no. 4, pp: 40 1–416, 2007.
53. R. Das, Chip-less RFID – The end game, IDTechEx internet article, Feb. 2006 http://www.idtechex.com/products/en/articles/00000435.asp (accessed February 2006).
54. M. Glickstein, Firewall protection for paper documents, RFID Journal internet article, Feb. 2004, http://www.rfidjournal.com/article/articleprint/790/-1/1/ (accessed February 2006).
55. J. Collins, RFID fibers for secure applications, RFID Journal internet article, March 2004. http://www.rfidjournal.com/article/articleprint/845/-1/1/ (accessed April 2006).
56. K. C. Jones, "Invisible tattoo ink for chipless RFID safe, company says", EE Times white paper, October 2007 http://eetimes.eu/industrial/196900063 (accessed June 2009).
57. J. Dowe, "SOMARK's chipless RFID ink tattoo field demo brings the company closer to launch", MoreRFID internet article, Feb. 2008. http://www.morerfid.com (accessed June 2009).
58. Somark Innovations, "Platform technology capabilities", white paper, 2007. http://somarkinnovations.com/technology/ (accessed June 2009).

59. I. Jalaly, I. D. Robertson, "RF barcodes using multiple frequency bands", *IEEE MTT-S International Microwave Symposium Digest 2005*, pp: 4–7, Long beach, USA, June 2005.
60. J. McVay, A. Hoorfar, N. Engheta, "Theory and experiments on Peano and Hilbert curve RFID tags", *Proceedings of the Wireless Sensing and Processing*, vol. 6248, pp: 624808, San Diego, USA, Aug. 2006.
61. Tagsense, Inc., "Chipless RFID products", data sheet, http://www.tagsense.com/ingles/products/product_chipless.html (accessed October 2006).
62. Mark Palmer, "Build an effective RFID architecture", RFID Journal internet article, Feb 2004. Available: http://www.rfidjournal.com/article/articleview/781/1/82/ (accessed June 2007).
63. D. M. Pozar, "Microwave engineering – 3rd Edition", John Wiley & Sons, Inc., 2005.
64. P. Salonen, L. Sydanheimo, "A 2.45 GHz digital beam-forming antenna for RFID reader, *IEEE 55th Vehicular Technology Conference*, vol. 4, pp. 1766–1770, Birmingham, Alabama, USA, May 2002.
65. S. K. Padhi, N. C. Karmakar, C. L. Law, S. Aditya, "A dual polarized aperture coupled microstrip patch antenna with high isolation for RFID applications, *2001. IEEE Antennas and Propagation Society International Symposium*, vol. 2, pp. 2–5, Boston, USA, July 2001.
66. G. Marocco, "Gain-optimized self-resonant meander line antennas for RFID applications", *Antennas and Wireless Propagation Letters*, vol. 2, pp. 302–305, 2003.
67. R. Hornung, "ARLON FoamClad based microstrip patch antennas and arrays for RFID readers", ARLON internet white paper, August 2005. Available: http://www.arlonmed.com/Foamclad%20for%20RFID.pdf#search='microstrip%20antennaRFID%20reader' (accessed May 2006).
68. CHIPSILICON Pty. Ltd, Product information internet site, 2006. http://info.chipsilicon.com:8080/chipsilicon/products/ (accessed May 2006).
69. Omron Canada Inc., "Microwave RFID system V690 data sheet", product data sheet, 2005. Available: http://www.omron.com (accessed March 2006).
70. Alien Technology Corporation, BAP ALR-2850 reader data sheet, 2005. Available: http://www.alientechnology.com/products/rfid_readers.php (accessed February 2006).
71. AVANTE International Inc., "AVANTE Handheld RFID Reader product info", web site, 2006. Available: http://www.avantetech.com (accessed March 2006).
72. Dallas Semiconductor MAXIM, "Application note 83: Fundamentals of RS-232 Serial Communications", application note, March 2001. Available: www.maxim-dallas.com (accessed March 2006).
73. Feig Electronic Gmbh, "Product information internet site", web site, 2006. Available: http://www.feig.de (accessed March 2006).
74. Idesco, "IR8000 reader data sheet", product data sheet, May 2005. Available: http://www.idesco.fi (accessed March 2006).
75. Smart Code Corporation, "Product information internet site", web site, 2006. Available: http://www.smartcodecorp.com/products/rfid_readers.asp (accessed April 2006).
76. Samsys Technologies Inc., "UHF reader MP9320 v2.8e data sheet", product data sheet, 2005. Available: http://ww.samsys.com (accessed March 2006).
77. Thingmagic Corporation, "Product information internet site", web site, 2006. Available: http://www.thingmagic.com/html/prod.htm (accessed February 2006).
78. Symbol Corporation, "XR480 RFID reader data sheet", product data sheet, 2006. Available: http://www.symbol.com/XR480 (accessed March 2006).
79. TAGSYS Corporation, "Product information internet site", web site, 2006. Available: www.tagsys.com/html/rfid-53.html (accessed March 2006).
80. Tagmaster, "LR-3 reader product data sheet", product data sheet, 2006. Available: http://www.tagmaster.com (accessed March 2006).
81. RF Code, Inc, "Mantis II Active RFID reader data sheet", product data sheet, 2006. Available: http://www.rfcode.com/433mantis_reader.asp (accessed April 2006).
82. Identec Solutions, Inc., "i-CARD CF Mobile reader data sheet", product data sheet, 2006. Available: http://www.identecsolutions.com/i-cardCF.asp (accessed March 2006).
83. SAVI Technology Inc., "SAVI Mobile reader SMR-650 data sheet", product data sheet, 2005. Available: http: //www.savi.com/products/ds.smr650.pdf (accessed March 2006).

84. V. Naware, G. Mergen, L. Tong, "Stability and delay of finite-user Slotted ALOHA with multipacket reception", *IEEE Transactions of Information Theory*, vol. 51, issue 7, pp: 2636–2656, July 2005.
85. C. Nak-Gwon, L. Hyuek-Jae, L. Sang-Hoon, K. Seong-Jeen, "Design of a 13.56MHz RFID system", *ICACT 2006 The 8th International Conference of Advanced Communication Technology*, vol. 1, Phoenix Park, Korea, February 2006.
86. ActiveWave Inc., "Product information data internet site", web site, 2006. Available: http://www.activewaveinc.com/products_rfid_readers.html (accessed March 2006).
87. R. Page, "A low power RF ID transponder", *Grand Prize Winner 1993 RF Design Awards Contest*, Wenzel Associates, 1993. Available: http://www.wnzel.com/pdffiles1/pdfs/rays.pdf (accessed February 2006).
88. BALOGH RFID, "HYPER X LMB-6012 RFID reader data sheet", product data sheet, January 2003. Available: http://www.balogh.cc/HyperX/Support/PDF/LMB6012-6013.pdf (accessed March 2006).
89. RFSAW, Inc., "RFSAW RFID system data sheet", product data sheet, 2003. Available: www.rfsaw.com (accessed April 2006).
90. Impinj, "The Gen 2 story: Charting the path to RFID that just works", internet white paper, 2005 Available: http://www.impinj.com/files/MR_GP_ED_00001_Gen2Story.pdf (accessed June 2007).
91. EPC Global, "EPC radio-frequency identity protocols Class-1, Generation-2 UHF RFID protocol for communications at 860 MHz-960 MHz – version 1.0.9", Internet article, EPCglobal Inc, Jan. 2005 Available: http://www.epcglobalinc.org/standards/Class_1_Generation_2_UHF_Air_Interface_Protocol_Standard_Version_1.0.9.pdf (accessed June 2007).
92. Alien Technology, "EPCglobal Class 1 Gen 2 RFID specification", internet whitepaper, 2005. Available: http://www.alientechnology.com/docs/AT_wp_EPCGlobal_WEB.pdf (accessed June 2007).
93. EE Times Asia, "RFID solution enables single universal RFID reader", internet article, January 2006. Available: http://www.eetasia.com/ART_8800403199__NP_798750a5.HTM (accessed June 2007).
94. P. Salonen, M. Keskilammi, L. Sydanheimo, M. Kivikoski, "An intelligent 2.45 GHz beam-scanning array for modern RFID reader", *IEEE International Conference on Phased Array Systems and Technology*, pp. 407–410, Dana Point, USA, May 2000.
95. T. Nakamura, J. Seddon, "Omron develops world's first antenna technology that boosts UHF RFID tag read performance", Omron Corporation press releases, 2006. Available: http://www.omron.com/news/n_270306.htmlC (accessed June 2007).
96. Profibus-DeviceNet-Ethernet IP-Modbus-Remote I/O, http://www.rfidinc.com (accessed June 2007).
97. Texas Instruments, Inc., "Series 2000 Micro reader data sheet", product data sheet, 2002. Available: http://www.ti.com/rfid/docs/manuals/pdfSpecs/RI-STU-MRD1.pdf (accessed June 2006).
98. Sentinel ID Systems, Inc., "Product information internet site", web site, 2006. Available: http://www.sentinelid.com/products.html (accessed March 2006).
99. M. Reynolds, C. Weigand, "Design considerations for embedded software-defined RFID readers", internet white paper, August 2005. Av ailable: http://www.thingmagic.com/html/508EWTF2.pdf#search='508EWTF2' (accessed April 2006).
100. Electronic Engineering Times, "Anadigm rolls universal RFID reader", internet site, June 2005. Available: http://www.eetasia.com/ART_8800368056_499491_a7168a56.HTM (accessed March 2006).
101. Anadigm Inc., "RangeMaster solution for RFID tag readers", Anadigm product data sheet, 2005. Available: http://www.anadigm.com/_doc/GF041100-0001.pdf#search='anadigmrangemaster' (accessed March 2006).
102. J. Joubert, "Spiral microstrip resonator for narrow-stopband filters", *Proc. IEE – Microwave, Antennas, Propag.*, vol. 150, no. 6, pp: 493–496, Dec 2003.
103. N. C. Karmakar, S. M. Roy, I. Balbin, "Quasi-static modelling of defected ground structure", *IEEE Transactions on Microwave Theory and Techniques*, vol. 54, no. 5, pp: 2160–2168, May 2006.

104. S. M. Roy, N. C. Karmakar, "Analysis of different DGS configurations in terms of microstrip discontinuities", *International Conference on Electrical and Computer Engineering*, pp: 544–547, Dhaka, Bangladesh, December 2006.
105. M. N. Mollah, N. C. Karmakar, "A novel hybrid defected ground structure as low pass filter", *IEEE Antennas and Propagation Society International Symposium 2004*, vol. 4, pp: 3581–3584, Monterey, USA, June 2004.
106. D. Woo, T. Lee, "Suppression of harmonics in Wilkinson power divider using dual-band rejection by asymmetric DGS", *IEEE Transactions on Microwave Theory and Techniques*, vol. 53, no. 6, pp: 2139–2144, June 2005.
107. J. Lim, J. Park, Y. Lee, D. Ahn, S. Nam, "Application of defected ground structure in reducing the size of amplifiers", *IEEE Microwave and Wireless Component Letters*, vol. 12, no. 7, pp: 261–263, July 2002.
108. D. Woo, J. Lee, T. Lee, "Multi-band rejection DGS with improved slow-wave effect", *38th European Microwave Conference*, pp: 1342–1345, Amsterdam, Netherlands, October 2008.
109. Y. Lee, J. Sun, "CPW band-stop filter with tapered-shaped defected ground structure", *Asia-Pacific Microwave Conference 2007 APMC 2007*, pp: 1–4, Bangkok, Thailand, December 2007.
110. M. Joung, J. Park, H. Kim, "A novel modeling method for defected ground structure using adaptive frequency sampling and its application to microwave oscillator design", *IEEE Transactions on Magnetics*, vol. 41, no. 5, pp: 1656–1659, May 2005.
111. A. M. E. Safwat, F. Podevin, P. Ferrari, A. Vilcot, "Tunable bandstop defected ground structure resonator using reconfigurable dumbbell-shaped coplanar waveguide", *IEEE Transactions on Microwave Theory and Techniques*, vol. 54, no. 9, pp: 3559–3564, Sept. 2006.
112. Y. Lee, K. Nae, E. Lee, H. Ron, J. Jung, Y. Han, J. Jung, S. Baek, M. Jung, H. Cho, H. Oh, Y. Seong, J. Park, "ASK modulator for RFID system using a novel variable DGS resonator", 37th *European Microwave Conference EuMC 2007*, pp: 381–384, Munich, Germany, October 2007.
113. K. Ma, K. Yeo, J. Ma, M. Do, "An ultra-compact planar bandpass filter with open-ground spiral for wireless application", *IEEE Transactions on Advanced Packaging*, vol. 31, no. 2, pp: 285–291, May 2008.
114. H. Lim, J. Lee, S. Lim, D. Shin, N. Myung, "A novel compact microstrip bandstop filter based on spiral resonators", *Asia-Pacific Microwave Conference APMC 2007*, pp: 1–4, Bangkok, Thailand, Dec. 2007.
115. Y. Lee, J. Lim, C. Kim, D. Ahn, S. Nam, "A compact-size microstrip spiral resonator and its application to microwave resonator", *IEEE Microwave and Wireless Components Letters*, vol. 12, no. 10, pp: 375–377, Oct. 2002.
116. K. Yoon, H. Lee, J. Park, J. Lee, "Design of high-Q resonator for satellite broadcasting application", *IEEE Antennas and Propagation Society International Symposium AP-S 2008*, pp: 1–4, San Diego, USA July 2008.
117. Z. Jiang, P. S. Excell, Z. M. Hejazi, "Calculation of distributed capacitance of spiral resonators", *IEEE Transactions on Microwave Theory and Techniques,* vol. 45, no. 1, pp: 139–142, January 1997.
118. Z. Hejazi, P. S. Excell, Z. Jiang, "Accurate distributed inductance of spiral resonators", *IEEE Microwave and Guided Wave Letters*, vol. 8, no. 4, pp: 164–166, April 1998.
119. A. P. S. Khanna, Y Garault, "Determination of loaded, unloaded and external quality factors of a dielectric resonator coupled to a microstrip line", *IEEE Transactions of Microwave Theory and Techniques*, vol. MTT-31, pp: 261–264, March 1983.
120. C. P. Wen, "Coplanar waveguide: a surface strip transmission line suitable for nonreciprocal gyromagnetic device applications", *IEEE Transactions on Microwave Theory and Techniques*, vol. 17, no. 12, pp: 1087–1090, December 1969.
121. FCC, First Report and Order 02–48, February 2002.
122. K. Y. Yazdandoost, R. Kohno, "Ultra wideband antenna", *IEEE Communication Magazine*, vol. 42, no. 6, pp: S29–S32, 2004.
123. X. Chen, P. J. Massey, "Operating principles and features of UWB monopoles and dipoles", *The Institution of Engineering and Technology Seminar on Ultra Wideband Systems, Technologies and Applications 2006*, pp: 131–152, Stevenage, UK, April 2006.

124. X. Chen, L. Guo, J. Liang, C. Parini, "On the performance of UWB monopole antennas", *IEEE International Conference on Ultra-Wideband ICUWB 2007*, pp: 210–213, Singapore, Sep. 2007.
125. X. Chen, J. Liang, P. Li, L. Guo, C. Chiau, C. Parini, "Planar UWB monopole antennas", *Asia-Pacific Microwave Conference APMC 2005*, vol. 1, pp: 1–4, Suzhou, China, May 2005.
126. A. Alipour, H. R. Hassani, "A novel omni-directional UWB monopole antenna", *IEEE Transactions on Antennas and Propagation*, vol. 56, no. 12, pp: 3854–3857, Dec. 2008.
127. Q. Wu, J. Ronghong, J. Geng, M. Ding, "Printed omni-directional UWB monopole antenna with very compact size", *IEEE Transactions on Antennas and Propagation*, vol. 56, no. 3, pp: 896–899, March 2008.
128. G. Quintero, A. K. Skrivervik, "Analysis of planar UWB elliptical dipoles fed by a coplanar stripline", *IEEE International Conference on Ultra-Wideband ICUWB 2008*, vol. 1, pp: 113–116, Hannover, Germany, Sep. 2008.
129. M. Mudroch, P. Cerny, P. Hazdra, M. Mazanek, "UWB dipole antenna optimization with neural network tuned algorithm", *3rd European Conference on Antennas and Propagation EUCAP 2009*, pp: 1491–1494, Berlin, Germany, March 2009.
130. G. Whyte, F. Darbari, I. McGregor, I. Glover, I. Thayne, "Different feeding geometries for planar elliptical UWB dipoles, and the excitation of leakage current", *38th European Microwave Conference EuMC 2008*, pp: 1382–1385, Amsterdam, Netherlands, Oct. 2008.
131. V. Gonzalez-Posadas, C. Martin-Pascual, J. L. Jimenez-Martin, D. Segovia-Vargas, "Lumpedelement balun for UHF UWB printed balanced antennas", *IEEE Transactions on Antennas and Propagation*, vol. 56, no. 7, pp: 2102–2107, July, 2008.
132. E. E. Angelopoulos, A. Z. Anastopoulos, D. I. Kaklamani, A. A. Alexandridis, F. Lazarakis, K. Dangakis, "Circular and elliptical CPW-Fed slot and microstri-fed antennas for ultrawideband applications", *IEEE Antennas and Wireless Propagation Letters*, vol. 5, no. 1, pp: 294–297, December 2006.
133. Y. Kuo, K. Wong, "Printed double-T monopole antenna for 2.4/5.2 GHz dual-band WLAN operations", *IEEE Transactions on Antennas and Propagation*, vol. 51, no. 9, pp: 2187–2191, Sep. 2003.
134. J. Jung, K. Seol, W. Choi, J. Choi, "Wideband monopole antenna for various mobile communication applications", *Electronic Letters*, vol. 41, no. 24, pp: 1313–1314, Nov. 2005.
135. K. Chung, J. Kim, J. Choi, "Wideband microstrip-fed monopole antenna having frequency bandnotch function", *IEEE Microwave and Wireless Components Letters*, vol. 15, no. 11, pp: 766–768.
136. Y. J. Chon, K. H. Kim, S. H. Hwang, S. Park, "A miniature UWB planar monopole antenna with 5GHz band-rejection filter", *35th European Microwave Conference,* pp: 1911–1914, October 2005, Paris, France.
137. H. Chen, H. Chen, "A CPW-fed dual-frequency monopole antenna", *IEEE Transactions on Antennas and Propagation*, vol. 52, no. 4, pp: 978–982, April 2004.
138. J. -Y. Chan, T. -M. Kuo, "CPW-fed wideband planar monopole antenna for operations in DCS, PCS, 3G, and Bluetooth bands", *Electronics Letters*, vol. 41, no. 18, pp: 991–993, September 2005.
139. Y. Kim, D.-H. Kwon, "CPW-fed planar ultra wideband antenna having a frequency band notch function", *Electronic Letters*, vol. 40, no. 7, pp: 403–405, April 2004.
140. W. Wang, S. S. Zhong, S. Chen, "A novel wideband coplanar-fed monopole antenna", *Microwave and Optical Technology Letters*, vol. 43, no. 1, pp: 50–52, October 2004.
141. S. H. Lee, J. K. Park, J. N. Lee, "A novel CPW-fed ultra-wideband antenna design", *Microwave and Optical Technology Letters*, vol. 44, no. 5, pp: 393–396, March 2005.
142. C. A. Balanis, "Antenna theory: analysis and design", 2nd Edition, John Wiley & Sons, New York, 1982.
143. K. V. S. Rao, P. V. Nikitin and S. M. Lam, "Antenna design for UHF RFID tags: a review and a practical application", *IEEE Transactions on Antennas and Propagation*, vol. 53, issue 12, pp: 3870–3876, Dec. 2005.

144. P. V. Nikitin, K. V. S. Rao, "Reply to "Comments on 'Antenna design for UHF RFID tags: A review and a practical application'"," *IEEE Transactions on Antennas and Propagation*, vol. 54, no. 6, pp: 1906–1907, June 2006.
145. RFSAW Inc, "The global SAW tag – a new technical approach to RFID", internet white paper, 2004. Available: http://www.rfsaw.com/pdfs/SAW%20RFID%20Whitepaper.pdf (accessed: October 2009).
146. RFSAW Inc, "Model 501 SAW RFID Reader System", product data sheet, 2004. Available: http://www.rfsaw.com/pdfs/Data_501_4.pdf (Accessed October 2009).
147. WJCommunications Inc, "A compact RFID reader platform for UHF and microwave applications", *Microwave Journal*, vol. 47, no. 5, pp: 42–48, April 2004.
148. F. Mohd-Yasin, M. K. Khaw, M. B. I. Reaz, "Radio frequency identification: Evolution of reader and antenna circuit design", *Microwave Journal*, vol. 49, no. 5, pp: 89–95, May 2007.
149. M. J. Uddin, A. N. Nordin, M. I. Ibrahimy, M. B. I. Reaz, T. Z. A. Zulkifli, M. A. Hasan, "Design and simulation of RF-CMOS spiral inductors for ISM band RFID reader circuits", *IEEE Workshop on Microelectronics and Electron Devices 2009*, pp: 1–4, April 2009.
150. G. K. Balachandran, R. E. Barnett, "A high dynamic range ASK demodulator for passive UHF RFID with automatic over-voltage protection and detection threshold adjustment", *IEEE Custom Integrated Circuits Conference CICC 2009*, pp: 383–386, San Jose, USA, Sept. 2009.
151. G. De Vita, G. Iannaccone, "Ultra low power RF section of a passive microwave RFID transponder in 0.35μm BiCMOS", *IEEE International Symposium on Circuits and Systems ISCAS 2005*, vol. 5, pp: 5075–5078, Kobe, Japan, May 2005.
152. I. Mayordomo, R. Berenguer, I. Fernandez, I. Gutierrez, W. Strauss, J. Bernhard, "Simulation and measurement of a long-range passive RFID system focused on reader architecture and backscattering communication", *38th European Microwave Conference EuMC 2008*, pp: 1058–1061, Amsterdam, Netherlands, October 2009.
153. T. Osinski, "Why buy EPCglobal-Certified products", *RFID Journal internet article*, April 2009, Available: http://www.rfidjournal.com/article/view/4781/1 (Accessed: October 2009).
154. IDX Inc, "Cashless payment system", Product data sheet, August 2009, Available: http://www.idxinc.com/Wickets/downloads/Casino%20Account%20Based%20Wickets%20Brochure.pdf (accessed: October 2009).
155. Z. N. Chen, X. Qing, H. L. Chung, "A universal UHF RFID reader antenna", *IEEE Transactions on Microwave Theory and* Techniques, vol. 57, no. 5, pp: 1275–1282, May 2009.
156. Tropos Networks Inc, "Receiver sensitivity – a technology brief", internet article, July 2007, Available: http://www.tropos.com/pdf/technology_briefs/tropos_techbrief_rx_sensitivity.pdf (accessed: October 2009).
157. W. Lim, S. Park, W. Son, M. Lee, J. Yu, "RFID reader front-end having robust Tx leakage canceller for load variation", *IEEE Transactions on Microwave Theory and Techniques*, vol. 57, no. 5, pp: 1348–1355, May 2009.
158. K. Penttil, L. Sydnheimo, M. Kivikovski, "Implementation of Tx/Rx isolation in an RFID reader", *Int. J. Radio. Freq. Identification Technol. Appl.*, vol. 1, no. 1, pp: 74–79, 2006.
159. S. L. Karode, V. F. Fusco, "Feedforward embedding circulator enhancement in transmit/receive applications", *IEEE Microw. Wireless Compon. Lett.*, vol. 8, no. 1, pp: 33–34, Jan. 1998.
160. J. H. Jung, J. H. Kim, S. M. Kim, K. C. Lee, "New circulator structure with high isolation for time division duplexing radio systems", *IEEE Veh. Technol. Conf.*, vol. 4, pp: 2766–2769, Stockholm, Sweden, Sep. 2005.
161. K. Choi, Y. Eo, S. Jung, I. Kwon, H. Lee, Y. Kim, "A fully integrated CMOS RF transmitter for UHF mobile RFID reader applications", *37th European Microwave Conference EuMC 2007*, pp: 1648–1651, Munich, Germany, Oct. 2007.
162. W. Lim, J. Yu, "Balanced circulator structure with enhanced isolation characteristics", *Microw. Opt. technol. Lett.*, vol. 50, no. 9, pp: 2389–2391, Sep. 2008.
163. C. Hartmann, P. Hartmann, P. Brown, J. Bellamy, L. Claiborne, W. Bonner, "Anti-collision methods for global SAW RFID tag system", *IEEE Ultrasonics Symposium*, vol. 2, pp: 805–808, Montreal, Canada, August 2004.

GPSR Compliance

The European Union's (EU) General Product Safety Regulation (GPSR) is a set of rules that requires consumer products to be safe and our obligations to ensure this.

If you have any concerns about our products, you can contact us on ProductSafety@springernature.com

In case Publisher is established outside the EU, the EU authorized representative is:

Springer Nature Customer Service Center GmbH
Europaplatz 3
69115 Heidelberg, Germany

Batch number: 10614746

Printed by Printforce, the Netherlands